VOLUME ONE HUNDRED AND NINETEEN

ADVANCES IN
COMPUTERS

VOLUME ONE HUNDRED AND NINETEEN

ADVANCES IN
COMPUTERS

Edited by

ALI R. HURSON
Missouri University of Science and Technology
Rolla, MO, United States

ELSEVIER

ACADEMIC PRESS

An imprint of Elsevier

Academic Press is an imprint of Elsevier
50 Hampshire Street, 5th Floor, Cambridge, MA 02139, United States
525 B Street, Suite 1650, San Diego, CA 92101, United States
The Boulevard, Langford Lane, Kidlington, Oxford OX5 1GB, United Kingdom
125 London Wall, London, EC2Y 5AS, United Kingdom

First edition 2020

ISBN: 978-0-12-820325-5
ISSN: 0065-2458

For information on all Academic Press publications
visit our website at https://www.elsevier.com/books-and-journals

Publisher: Zoe Kruze
Editorial Project Manager: Hilal Johnson
Production Project Manager: James Selvam
Cover Designer: Greg Harris

Typeset by SPi Global, India

Working together
to grow libraries in
developing countries

www.elsevier.com • www.bookaid.org

Contents

4. Eyeing the patterns: Data visualization using doubly-seriated color heatmaps   **121**

Matthew Lane, Alberto Maiocco, Sanjiv K. Bhatia, and Sharlee Climer

5. Eigenvideo for video indexing   **157**

Nora Alosily and Sanjiv K. Bhatia

Preface

Traditionally, *Advances in Computers*, the oldest series to chronicle of the rapid evolution of computing, annually publishers several volumes, each one typically comprised of four to eight chapters, describing new developments in the theory and applications of computing.

The 119th volume is an eclectic volume inspired by recent several issues of interest in research and development in computer science/computer engineering, and several closely related disciplines. The volume is a collection of five chapters as follows.

Chapter 1 entitled, "Fast execution of RDF queries using Apache Hadoop," by Mazumdar and Scionti from Italy is inspired by the ability of Ma-Reduce in handling the so-called Big Data. The article presents a framework for fast evaluation of complex queries in SPARQL. SPARQL is a query language aimed at retrieving and manipulating data stored in RDF format data that supports Big Data applications as well. The chapter also addresses the implemented of the proposed query engine on Hadoop framework. In addition, as discussed in the article, parallel bloom filter is used to preprocess the input data with the goal of further improvement of the query processing. Finally, to prove the concept, the query engine is tested and results are analyzed.

Chapter 2 articulates the demand on high performance computing and high cost of power consumption of high-end computational platform. Hajiamini and Shirazi in "A study of DVFS methodologies for multicore systems with islanding feature" proposing the application of dynamic voltage and frequency scaling (DVFS) technique for scaling voltage/frequency (V/F) levels of cores based on their time-varying workloads during application runtime. The article further investigates improving multicore platforms' energy efficiency with DVFS with the goal of minimizing execution time while maintaining the energy consumption below a user-defined energy budget or minimizing the energy consumption while maintaining a desirable execution time defined by the user. The proposed optimization technique is evaluated on multiple applications with varying computational characteristics.

Laxmi et al. in Chapter 3 entitled, "Effectiveness of state-of-the-art dynamic analysis techniques in identifying diverse Android malware and future enhancements" articulates the popularity of Android and as a consequence its vulnerability to malware writers by creating inroads into Android devices through malicious apps. The malicious apps may access and leak personal and sensitive information leading to fraud, financial loss, and …. This chapter reviews and analyzes the effectiveness of various state-of-the-art dynamic analysis techniques against antidetection techniques and attempts to highlight issues and challenges that require the attention of research community in creating a more reliable environment.

Chapter 4 entitled, "Eyeing the patterns: Data visualization using doubly-seriated color heatmaps" by Lane et al., focuses on heatmap as a visualization technique to facilitate pattern extraction. In heatmaps is a two-dimensional space where a row represents an object and a column represents a condition or other property. The color of a cell indicates the associated data value. However, observations of randomly ordered data are rarely enlightening and rearrangement of rows and columns into clusters of similar data is of great value. This chapter, briefly overviews the history of heatmaps, analyzes various aspects of data preprocessing, and examines several algorithms such as Bond Energy Algorithm, the Traveling Salesman Problem Model, $TSP + k$, and Hierarchical Clustering for reordering data rows and columns for the purpose of clustering similar data. Finally, the chapter concludes with thoughts for potential future directions of heatmap research.

Finally, Chapter 5 concentrates on video databases and video indexing. Alosily and Bhatia in "Eigenvideo for video indexing" propose the application of principal component analysis to reduce the dimensionality in large-scale video dataset. This analysis determines the best eigenvectors of the dataset that have the smallest eigenvalues. The resultant eigenvectors encapsulate the true dimensionality of video datasets. They present the algorithm that starts by building a representative feature vector for each scene in the dataset. The representative feature vectors then determine the eigenvectors that represent the whole dataset. The proposed algorithm is analyzed for the proof of concept.

I hope that readers find this volume of interest, and useful for teaching, research, and other professional activities. I welcome feedback on the volume, as well as suggestions for topics of future volumes.

ALI R. HURSON

Missouri University of Science and Technology

Rolla, MO, United States

CHAPTER ONE

Fast execution of RDF queries using Apache Hadoop

Somnath Mazumdar[a], Alberto Scionti[b]
[a]Department of Information Engineering and Mathematics, Università di Siena, Siena, Italy
[b]Istituto Superiore Mario Boella (ISMB), Turin, Italy

Contents

Abstract

Map-Reduce (MR) is a distributed programming framework which became very popular since its introduction, due to its ability to process massive data sets. MR provides a robust and straightforward mechanism to implement distributed applications without worrying much about many management aspects of parallel programming (e.g., instantiating jobs, data distribution, job synchronization). On the other hand, the Resource Description Framework (RDF) with its simplicity and flexibility, can represent semistructured and unstructured data which are very important for representing web-semantics. SPARQL

Advances in Computers, Volume 119
ISSN 0065-2458
https://doi.org/10.1016/bs.adcom.2020.03.001

is a query language aimed at retrieving and manipulating data stored in RDF format and also supports "Big Data" applications. In this book chapter, we present a framework designed to evaluate complex SPARQL queries fast. To improve the execution of SPARQL queries, we implemented the query engine on the Hadoop framework. The engine can handle large and complex queries involving multiple join variables while running on large RDF data sets. Further execution speedup is obtained by preprocessing the input data with parallel Bloom filters. The query engine has been tested on the SP^2 benchmark, and the results demonstrate the benefits of the design. In this case, the minimum query improvement is 5% while the maximum improvement has been achieved is 82%.

Acronym	Description
ACID	Atomicity consistency isolation and durability
API	Application program interface
ARQ	A SPARQL processor for Jena
BF	Bloom filter
BGP	Basic graph pattern
CDN	Content delivery network
DBaaS	Big data as a service
DC	Data center
HDFS	Hadoop distributed file system
IaaS	Infrastructure as a service
MR	MapReduce
N3	Notation 3
OWL	Web ontology language
PaaS	Platform as a service
PBF	Parallel Bloom filter
RDF	Resource description framework
RISC	Reduced instruction set computer
SaaS	Software as a service
SPARQL	SPARQL protocol and RDF query language
SP^2	A SPARQL performance benchmark
SQL	Structured query language
YARN	Yet another resource negotiator

1. Introduction

In recent times, web applications, as well as technfologies, became more complex and interactive. Modern web applications provide flexibility to the user to interact and collaborate with others over the Internet. Given the ever-growing amount of data generated and exchanged by the users, data centers (DCs) are demanded to expose large computational capabilities to process such data. The primary challenges for managing large data sets or so-called Big Data are well-known five "Vs": volume, velocity, variability, veracity, and value. In other words, the Big Data related solutions must be able to handle: unbounded size of data, an unpredictable growth of new data, development of different types of data, the integrity issue of the newly generated data, and to extract inherent knowledge trapped inside the data. It is worth to note that resource management and data sets management are two independent components in distributed computing, but for a higher degree of efficiency, they must work in tandem. In the Cloud computing model, the data are stored in different level of abstractions, so that an overall consistent data management is achieved. The lowest abstraction level is represented by the Infrastructure-as-a-Service (IaaS) level, where general data storage options are: content delivery networks (CDNs), object storage, and volume storage (i.e., virtual hard drives). The Platform-as-a-Service (PaaS) level provides more abstracted/higher view of storage options (such as Database-as-a-Service (DBaaS), Big Data-as-a-Service (BDaaS)). Finally, Software-as-a-Service (SaaS) based storage can be accessed via web-services. Storing and efficiently retrieving a huge amount of information is not trivial, especially when complex conditions must be analyzed to query large amount of data. The complexity of managing such data also grows when their sheer size is considered.

Nowadays, an enormous amount of valuable web information is stored in Resource Description Framework—RDF [1] data format. RDF is built on public standards and offered as a platform for taking an agile approach with large and dynamic aggregations of data that would not fit into predefined tables. A human can visualize the RDF information in a graphical form, as well as in a statement form. In fact, web knowledge can be easily expressed in the form of very simple statements such as constructs. In addition, RDF offers the following (i) integration of data from different sources without using custom programming; (ii) capability of re-using data; (iii) decentralization of the data ownership; and (iv) capability to infer useful information from the RDF data model, while not being tied to a proprietary data storage/representation technology (such as database dialects). To extract meaningful information from RDF, the most well-known query language is SPARQL Protocol and RDF Query Language (SPARQL) [2]. SPARQL is recommended by W3C[a] and is considered as an important semantic web technology.

To extract the required information from a large data storage, queries should be executed. *Join* is one of the most common operations in the query evaluation process. It combines records from two or more tables based on given condition(s). There are many well-known algorithms for joining records (extracted from multiple data sets). When the join operation starts to execute on huge (structured, semistructured, or unstructured) data, performance slows down due to a large number of records to be scanned to provide the correct result. MapReduce (MR) programming framework [3] helps to tackle such challenges. It has been primarily designed by Google to support its web indexing technologies (such as crawling web pages, creating inverted-indices or analyzing web server access logs to find the top requested pages). MR also finds its popularity in applications such as query and image processing. MR has the advantage not to make any distinction between structured and unstructured data. Thus making it suitable for manipulating data in a very efficient manner [4]. Works have also been done to process relational database queries on MR (e.g., [5, 6]).

Query execution involves operations that extract and combine records depending on user-given conditions. Among the various operations, *join* is one of the fundamental, and most time as well as memory consuming operations to implement in the MR framework [7]. Evaluating complex queries on semistructured data is not trivial. In this book chapter, we advocate for a fast distributed SPARQL engine to evaluate complex queries, also involving multiple shared join keys or variables. MR is the basis of the proposed query engine. There is always need for combining huge intermediate data sets while evaluating complex queries containing multiple join keys on huge input data sets. Our engine tackles this challenge by leveraging the capability of the MR framework to distribute the workload on available computational resources. The primary motivation of this work is to develop a multijoin variable supported query processor which can run on standard MR framework. Storing the triples are expensive [8]. So, unlike other works which use other technologies (such as HBase) as their storage [9–11], we

---

[a] https://www.w3.org/.

have used HDFS [12] as the underlying storage system. Existing works (such as [8, 13, 14]), focus on proposing relational query processors to store and manage the RDF graphs. However, we have used triples format for querying and processing the RDF data. For query plan generation, we have used the BGP (basic graph pattern, the basic representational feature of SPARQL) thus avoiding employing unnecessary technologies (comparing to [15–17]). To optimize the first (preprocessing) phase, we employed a parallel version of the Bloom filter (BF) [18]. Apart from the support of the query processor, we also have shown that query optimization can be improved from 5% to 82% using a parallel version of BF.

The rest of the chapter is organized as follows. Section 2 reports research works related to this domain; Section 3 provides an overview of the used technologies in the book chapter. In Section 4, we detail how the join operations are performed in the MR framework; while results are shown in Section 5. Finally, we conclude the presented chapter in Section 6.

2. Related work

In recent years, RDF data query optimization has received huge attention from the research community. Indeed, the availability of diverse as well as massive data sets generated from various sources (such as sensors, Internet of things (IoT)-based applications) demands systems able to store such a huge amount of data. Apart from that it also needs to process these data to quickly get an insight of the events/systems. Among the others, RDF is an important ingredient for managing semantics of such data [19, 20]. The relational databases can process RDF datasets. However, with the growing size of triplets (as demanded by RDF processing), processing operations quickly incur a large slowdown of the system. Several studies have been done on speeding up execution of RDF related queries [21, 22]. Among the others, some approaches investigated the use of greedy/heuristics [23] to quickly generate optimized query plans, others explored reduced instruction set computer (RISC)-based optimized engines to gain query execution performance [24]. We refer readers to (i) [25] that surveys multiple distributed RDF data processing systems and grouped them using multiple characteristics; (ii) [8], that focuses on relational query processors to store and query RDF data; (iii) [26] which discusses the parallel and geo-distributed processing framework based on Hadoop and Spark; and (iv) [27] surveys MR-based algorithms for managing RDF graphs. From the large set of the literature, below we have only considered the works that are primarily based on Hadoop MR and HDFS.

Hadoop and HDFS-based approach: S2X combines graph–parallel and data–parallel computation for SPARQL while processing RDF data sets on Hadoop [14]. In this work, the authors define a property graph to represent the RDF and also designed a vertex-centric algorithm for BGP matching. However, S2X suffers from high overheads during BGP matching. In [15], the authors propose a SPARQL-over-SQL-on-Hadoop approach to support interactive-time SPARQL query processing on Hadoop framework. In this work, RDF is stored in a columnar layout on HDFS and uses a parallel processing SQL query engine for executing queries on Hadoop. CliqueSquare is an RDF data management platform built on top of Hadoop [16]. It uses a greedy clique-based algorithm for producing query plans that minimize the number of MR stages. This work also proposes a data partitioning strategy to reduce the amount of data transferred through the network during the query processing. Zhang et al. [13] proposes another technique called Entity Aware Graph compREssion (EAGRE) to model RDF data on key-value data store by preserving both the semantic, as well as structural information. EAGRE uses graph partitioning to distribute the RDF data to the compute nodes and later builds an in-memory index, to process queries. The primary aim of this approach is to reduce the query processing time as well as the I/O cost during the query processing. The authors also used a document indexing technique and join preprocessing technique for further optimization. In another work [28], the authors propose an algorithm (via key manipulation technique) for processing star-join queries in a constant number of computation steps. Similar to us, it also employs BFs for reduction of I/O. In [29], the authors developed a framework based on Hadoop to store and retrieve a large number of RDF triples. The authors also describe a schema to store RDF data in HDFS as well as algorithms to generate the query processing plan with bounded execution time (based on a worst case scenario). This work is quite similar to ours, but it does not implement any mechanism to filter unwanted data. In [30], the authors propose an algorithm for SPARQL's BGP to minimize the number of iterations using traditional multiway joins and to select a "good" join key to avoid unnecessary iterations. *Specialized approach*: In [31] the authors propose a distributed Hadoop-based SPARQL query processor for large-scale RDF data implemented on top of Spark. It uses a relational schema for RDF to avoid "dangling tuples"; thus reducing the query input size to join operations and also the total execution time. While DREAM stores the RDF data set intact at each cluster machine and employs a query planner that partitions any given SPARQL queries [32]. It proposes a graph-based query planner with a cost model and also developed a rule-oriented query partitioning strategy for choosing a suitable number of machines for executing SPARQL queries. R_2RDF+ is another query execution engine that performs join operations based on multiway merge and sort-merge join algorithms together with a cost model [11]. It also proposes an indexing scheme for storing RDF data implemented in HBase storage. There also exist multiple works that are focused on improving the execution or query support toward SPARQL queries such as [33] which proposes an algorithm to improve the RDF reasoning in MR framework; while [10] proposes an HBase-based hybrid RDF storage and query schema to separate static data and dynamic data. Similar to our work, Yang et al. [9] develops a SPARQL query classification algorithm to determine the join order to get a minimum number of MR jobs needed to execute the query. Unlike our approach, these works have used HBase-based tables extensively.

Here, we develop an RDF query engine on the top of *vanilla* MR framework. The proposed Hadoop-based query engine can execute complex queries involving multiple join variables. The Hadoop-MR framework is used to speed up the query execution by distributing processing tasks. However, to support massive data processing, we have added a simple yet efficient BF based preprocessing stage which aims at removing unwanted input data. It reduces both the execution time, as well as the data communication I/Os. Furthermore, for better query processing, we also have exploited the BGP. Compared to other works, the proposed query engine is built upon the standard Hadoop-MR framework and uses BGP as well as BFs. The proposed approach is generic and scalable, thus easing the process of porting to the Cloud.

3. Background

At the basis of the proposed query engine, there are several technologies that we briefly summarize in the following.

3.1 Hadoop MapReduce framework

The primary aim of developing the MR framework was to index Google's web pages. MR offers several features that enable developers to design efficient data processing solutions (such as scalability, ease of job distribution, intracluster communication, task monitoring, fault tolerance, backup, locality optimization, and an efficient sorting technique). MR's open source version called Hadoop [34] and the Hadoop framework adopts a distributed file system, referred to as the Hadoop Distributed File System (HDFS) [35]. It is well suited for applications adhering to the write-once-read-many data model.

The framework resembles a traditional master–slave architecture. Among the cluster nodes, one is marked as master, while the others are used as workers (i.e., slave nodes). The role of the master node is to coordinate the activity of the slaves by distributing tasks belonging to submitted jobs. The master runs a job tracker and each slave runs a task tracker. Slaves are in charge of executing user-defined *map* and *reduce* operations. To correctly distribute the tasks among the slaves, the job tracker continuously tracks the number of free map-reduce slots of each slave node. Such slots indicate the number of map and reduce operations that a specific node can execute. When a job is submitted to the system, job tracker tries to find out empty slots on the same node that hosts the data or a machine in the same rack. Task tracker monitors these spawned processes and notifies the job tracker about the current status. Fig. 1 provides a graphical representation of the interactions among architectural elements being part of the Hadoop framework during the data processing. The Hadoop framework provides the following main functions:

- *Map* receives input data as key-value pair (i.e., (K_1, V_1)) defined to a type in a data domain, and generates a list of key-value pairs (i.e., $list(K_2, V_2)$) with a type in a different data domain. A formal representation of the function is as follows:

$$map(K_1, V_1) \rightarrow list(K_2, V_2).$$

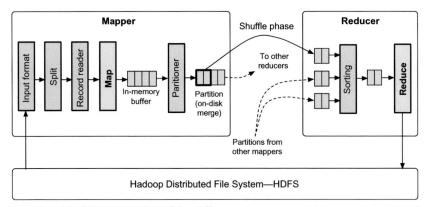

Fig. 1 Graphical representation of the different phases composing a generic MR task.

- *Reduce* receives a data pair composed by a key and a list of associated values (i.e., $(K_2, list(V_2))$) with a type in a data domain, and generates a collection of values (i.e., $list(V_3)$) in the same data domain. A formal representation of the function is as follows:

$$reduce(K_2, list(V_2)) \rightarrow list(V_3).$$

- *Sort* preprocesses reducer inputs by grouping data with the same key (i.e., it generates the pair $(K_i, list(V_i))$).
- *Shuffle* moves the sorted output of the mappers (i.e., the nodes that executed the map function) to the reducers (i.e., the nodes that will execute the reduce function). Interestingly, shuffle and sort phases occur concurrently while map-outputs are fetched and merged (i.e., the corresponding functions are pipelined).
- *Partition* determines in which reducer node a given (K_i, V_i) pair, corresponding to the output of the map phase, will be sent. Generally, the key K_i is used to derive the partition by applying a hash function, while the total number of partitions corresponds to the number of reduce tasks for the job.

Now, MR is also available in public Cloud VM instances. After the success of first Hadoop version (or "Hadoop 1.0"), a more modular Hadoop ("Hadoop 2.x") was developed (see Fig. 2). The new version introduced a new software layer called Yet Another Resource Negotiator (YARN),[b] which aims at dividing the functionalities of resource management and job scheduling (for both single or multiple job instances). However, the DC oriented resource management was not in the original scope of the YARN layer. To support DC-based infrastructures, a scalable resource manager called Mesos[c] started to be part of the framework (such as the Apache Myriad project[d]).

3.2 Resource description framework—RDF

RDF is considered one of the main pillars of the semantic web. RDF describes web resources using an XML-based syntax, as well as it provides a general method to decompose web knowledge into smaller pieces by following semantic rules. RDF concerns how information is represented both in a textual form using a specific syntax and through a graph.

---

[b] https://hadoop.apache.org/docs/current/hadoop-yarn/hadoop-yarn-site/YARN.html.
[c] http://mesos.apache.org/.
[d] http://myriad.incubator.apache.org/.

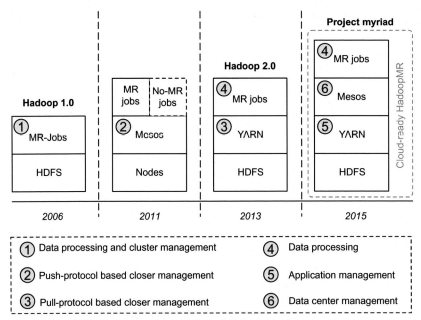

Fig. 2 The evolution of the Apache Hadoop-MR framework.

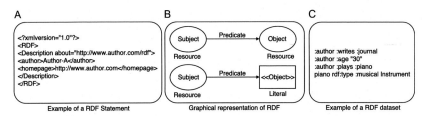

Fig. 3 Representational RDF models in graphical as well as in textual forms.

Such graph-based representation provides an insight on the data in a more structured way, and thus it better highlights relationships between facts. Fig. 3 shows an example of a simple fact represented using both RDF textual and graph-based model. Furthermore, the figure shows an example of an RDF data set.

Specifically, in Fig. 3A the RDF document describes its resources with properties, their associated values and the uniform resource identifiers (URIs) of the referenced resources. In fact, an RDF statement can be constructed by combining a resource (i.e., its identifier), a property, and the corresponding value. Alternatively, RDF statements consist of subjects, predicates, and objects. In that case, objects can also be text values called literals.

Fig. 3B depicts the general representation for an RDF statement of the form *"The author of http://www.author.com/rdf is Author-A."* using the graph-based modeling. In the graph-based representation, nodes are the subject and the object, while the edge is labeled with the predicate expressed by the statement. A triple composed of a subject, a predicate, and an object is also referred to as a fact. Finally, Fig. 3C shows an example of an RDF data set. An interesting aspect of the RDF representation of data is that we can infer new knowledge by simply linking together documents, taking into account their common vocabularies.

3.3 SPARQL

SPARQL is a standard way to query an RDF data set and is closely related to SPARQL Protocol and SPARQL Query Results XML format. In SPARQL, all IRIs (i.e., RDF URI references) are absolute and most forms of SPARQL query contain a set of triple patterns (i.e., the subject, the predicate or the object) known as BGP. BGPs are the basic building blocks of SPARQL query language and is a way of extracting subsets of nodes in an RDF graph, given a specific relationship. BGP contains the sequence of adjacent triple patterns in the query string. A BGP is matched against whatever is being queried and the results of matching are fed into the SPARQL query execution. Fig. 4 contains the example of a complex query where more than one atomic queries are linked together by a single common join variable. In the example "?x foaf:name ?name" and "?x foaf:mbox ?mbox" are both atomic queries, and they are linked together by a common join variable which is "?x". The given complex query will extract the value of "?name" and "?mbox" variables after the successful join operation. The queries getting more complex when multiple atomic queries are grouped together by multiple common join variables. The query contains "SELECT" and "WHERE" clauses. The former is used to identify the result variable and the latter generates the BGP to match the data. The execution

```
PREFIX foaf : <http://xmlns.com/foaf/0.1/>
SELECT ?name mbox
WHERE {
          ?x foaf:name ?name.
          ?x foaf:mbox ?mbox.
        }
```

Fig. 4 SPARQL query a complex query using QNames and containing a single join variable.

of SPARQL queries can be optimized by tuning the matching conditions for the BGP. To make the representation of the query more precise, we can use URIs or we can replace the URIs with their equivalent QNames. The query can be executed on data written in the Notation 3 (N3) format. N3 is a non-XML serialization of RDF models.

3.3.1 Joining SPARQL queries

Efficient and scalable query management using the RDF data is an open issue. SPARQL supports complex query processing involving single or multiple join variables. To execute any complex query, first, the query engine should perform BGP matching on the input graph and return valid results. BGP-based approach is a well-accepted approach [7]. There are two ways to perform join operation: (i) the two-way join, where join operation uses only two input files; (ii) the multiway join, which can handle multiple input files as join input. In SPARQL, conjunctive queries are expressed by using common variables across sets of BGPs. The detailed explanation can be found in Section 4.

3.4 Apache Jena framework

Jena is a popular and standard query execution environment offering a query language and a Web-API for SPARQL. It also provides a rich API for manipulating RDF graphs [36]. Jena Semantic Web Framework[e] is a rule-based inference engine for SPARQL query processing. It also provides an RDF API and storage strategies for RDF triples. Jena is mainly based on Java libraries to handle various semantic technologies (such as RDF, OWL (web ontology language) and SPARQL). RDF API in Jena is used to extract data from (or write them in) RDF, N3, and N-Triples formats. Jena's API has been primarily derived from SiRPAC API. Jena also has multiple internal reasoners, while the API allows to parse and search for RDF models. Jena supports two persistent storage mechanisms: TDB and SDB. TDB is a high-performance, native persistence engine using custom indexing and storage format, while SDB is a persistence layer that uses a more traditional SQL database supporting full ACID (i.e., atomicity, consistency, isolation and durability) properties. ARQ[f] is the query engine implementation (i.e., SPARQL processor) for Jena.

---

[e] https://jena.apache.org.
[f] https://jena.apache.org/documentation/query/index.html.

3.5 Bloom filter

Bloom filter (BF) is a probabilistic data structure which was developed in 1970 by Bloom. Initially, it was proposed for use in the web context by Marais et al. [37] as a mechanism for identifying which pages have associated comments stored within a common knowledge server. BFs offer data structures which are used to support membership queries (i.e., testing whether an element is a member of a given set or not). In fact, BF has the property of never returning any random data if the data are not in the set (i.e., input data are not a member of the given set).

Generally, a set $S = \{x_1, x_2, ..., x_n\} \in U$, where U is the universal set with cardinality $|U| = N$, can be stored in a sorted array using $n \cdot log(N)$ bits of memory. Querying the sorted array can require $O(log(n))$ operations. However, using a BF, the query time can be reduced to $O(1)$. BFs are characterized by the following properties:

- It provides a compact representation of very large sets.
- Time to test the membership of an element x_i to a set S (i.e., check if $x_i \in S$) is independent of the number of elements contained in S.
- False negatives are not possible (i.e., a member of the set S is reported as not contained in the set). However, false positives are possible (i.e., elements never inserted in the BF can be reported as members). Generally, this leads to a trade-off between space/time efficiency and the rate of false positives.
- Data cannot be deleted from the BF (i.e., this is true for canonical implementations). However, other newer versions of BF (e.g., Counting Bloom Filters [38]) support such operations.

Fig. 5 provides a visual representation of the operations performed by a BF (i.e., adding and verifying membership of a given element). BFs use a binary array v with a length of m bits to store membership information. All the elements of the array are initially set to zero. Registering a new member requires to define k independent hash functions (i.e., $h_1(\cdot), h_2(\cdot), ..., h_k(\cdot)$), which are used to map items $x \in S$ to random numbers uniformly distributed in the range $1, ..., m$. The values returned by the hash functions are thus interpreted as the position of the bits to set in the BF array. Once applied to an element $x \in S$, each hash function returns a value whose interpretation as an index allows to set a specific bit in the BF array. In other words, hash functions are applied sequentially, so that $v[h_i(x)] = 1, \forall i \in \{1, ..., k\}$. When querying the BF for the membership of an element $y \in U$, the value returned from the hash functions is checked. If any of the bits in the position provided by the hash functions is zero, the queried element is reported to be not a member

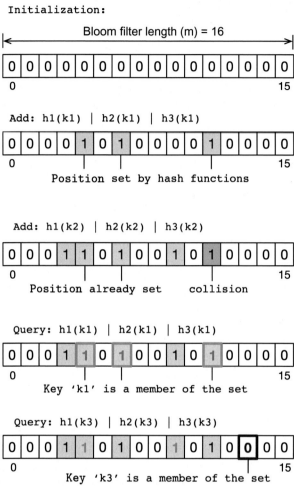

Fig. 5 Visual representation of adding and querying BF members.

of the set. Conversely, we conjecture that y is in the set S, although there is a certain probability that we are wrong (false positive). The parameters k and m should be chosen such that the probability of a false positive is acceptable. After inserting m keys into a table of size n, the probability that a particular bit is still zero is given by $\left(1 - \frac{1}{m}\right)^{k \cdot n}$. Eventually, the values returned by the hash functions can be combined (e.g., they can be XOR-ed) to set a single bit in the BF array, thus further reducing the probability of saturating the BF array. This approach allows supporting larger membership sets, although the probability of false positive can slightly increase.

Finally, it is worth noting that, since hash functions are independent, they can operate on the input keys in parallel. Every time a new element is registered in the BF array, the semantics of the set bits is maintained. In fact, two distinct hash functions may collide, thus setting the same bit position. Similarly, during membership verification, the BF array is accessed through a read operation, which does not modify the BF array content. In this perspective, a multithreaded BF implementation can provide better performance on large data sets.

3.5.1 Parallel Bloom filter (PBF)

We can parallelize BFs by distributing the hash functions' computations to multiple concurrent (parallel) threads. The large availability of multi-/many-core processors makes BF parallelization an interesting option to gain performance while verifying the membership of the input data. Aiming at increasing the system throughput, we modified the structure of the BF to speed up the membership verification. In fact, membership queries are the most common and used operations also for filtering input data in our MR-based query processor.

Fig. 6 shows the organization of the proposed multithreaded BF, referred to as parallel BF (PBF). Since the membership verification operation requires to access the BF storage to read values (such data structure can be safely shared among different concurrent threads) without incurring any race condition. Conversely, the input queue holding the batch of input data, to check for membership within the BF array, is split. As a parallelization

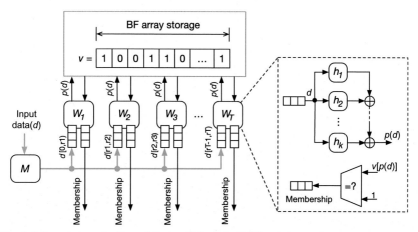

Fig. 6 Internal organization of the multithreaded BF implementation.

approach, we decided to replicate the operations to be performed for checking membership of input data, by splitting input batch data into subblocks, each of them assigned to a concurrent thread. Such approach also eases the parallel implementation, allowing to use common parallelization frameworks (such as OpenMP or multithreading libraries provided by programming languages). In fact, the inter-thread communication is unnecessary. Thus the implementation does not require to handle any specific communication pattern. Thus, a group of threads (*workers*—W_i) is instantiated, each receiving a fraction of the initial input batch to check.

Let R be the number of input data to check, and let T be the number of concurrent threads, each worker W_i will receive $r_{W_i} = \frac{R}{T}$ input data. To minimize the imbalance of workload assignment, the number of workers is generally set as a multiple of the number of available processing cores on the node which runs the PBF. This also eases the optimization of the memory management on processors with a complex cache memory hierarchy. During the membership verification process, each worker applies k independent hash functions, whose results are combined (i.e., the output of the hash functions are XOR-ed) to obtain a single index position in the BF array. Thus, given an input value d, the PBF computes the bit position $p(d)$ within the BF array as $p(d) = h_1(d) \oplus h_2(d) \oplus \dots \oplus h_k(d)$. A *master* thread (M) is responsible to fairly distribute the input requests to the worker input queues (blue arrows in Fig. 6). Finally, PBF query results are stored in multiple output queues (i.e., such queues can be implemented using simple output files, shared with other stages of the proposed framework).

4. Performing join operations in Hadoop

The main idea behind the proposed engine is to implement a mechanism that allows to speed up the execution of RDF queries. As discussed in Section 3.3, one of the most critical operations to perform is given by the *join* operator. To this end, in the following, our discussion will be focused on how to join and SELECT operations are implemented in the MR framework.

There are mainly two ways to implement join operations using Hadoop framework, which apply for the map phase and the reduce phase, respectively. They are referred to as *map-side join* and *reduce-side join*. In the former case, also named as *fragment-replicate join*, the MR framework holds one data set in memory by creating a hash table and joining on the other data set, record-by-record. The advantage of such join strategy is the minimal network and CPU overheads. In the latter case, data are grouped on the join key leveraging on the standard merge-sort function offered by the Hadoop framework. In such case, the framework emits a record for every pair of an element from the first data set and an element from the second data set. To avoid poor performance, the two data sets should be created in such way the one with fewest records per key comes first. In general, this approach provides an easy way to hold the records in memory and, if needed, it can go for the second pass. The two join strategies differ from each other in the amount of data that must be held in memory when operating. Considering the map-side join strategy, the smaller data set must fit in memory whereas, in the reduce-side join only each key group must fit in memory. It is worth to note that the number of workers for map-side jobs are always greater than the reduce-side jobs.

In general, a map-side join depends on the capability of exploiting specific characteristics of the data: (i) if the input is small enough to fit in memory and be replicated to every map; (ii) both inputs are already sorted on the same key, and (iii) both inputs are already partitioned into the same number of partitions using an identical hash function. If one of the above conditions are not true, then a preprocessing phase, which in turn is equivalent to a reduce-side join operation grouping keys and using Hadoop, must be added. For efficient execution, different matching and join algorithms for each phase may be applied, while each phase performs one MR job separately. Specifically, these jobs correspond to the input selection, the join operation, and the SELECT operation. However, to handle complex RDF queries, a simple and efficient implementation is required, although the number of iterative join operations depends on the number of common join variables.

Our current implementation (see Fig. 7) has four phases, which are executed as independent MR jobs (executing different matching and joining algorithms) in the cluster, and one data filtering stage (i.e., an intermediate job performed using BFs), which is executed on a dedicated node.

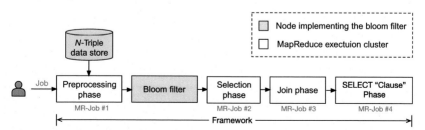

Fig. 7 The proposed framework implementation with the four MR jobs and the data filtering stage based on BFs.

- *Preprocessing phase* loads BFs. This phase will be executed whenever the input data set changes. It is an optimization process.
- *Selection phase* filters unnecessary input data and produces triples as output that matches at least one of the given query patterns. Here a query pattern is a single atomic query. The output of the selection phase has the key as the query pattern number and the value as the matching triples along with some extra information. This extra information is needed to perform the matching with the common join variables.
- *Join phase* loads the output of selection phase as the primary input. It generates RDF graph based on given complex query. A subgraph of an RDF graph is a subset of the triples in the graph. In this phase, one query execution plan is generated from RDF graph and runs iteratively (if it consists of multiple common join variables).
- *SELECT operation* has been used to project the required result from the join operation.

In the following, each of these separate phases (i.e., the MR jobs) is analyzed.

4.1 Preprocessing phase

Preprocessing phase has been implemented as the first MR job (running on a separated machine) to load the BFs. It is executed upon the change of the input data set. In this work, we structured the BF as *semantic BF*. Since it can adapt dynamically to the number of stored elements and also the *false positive probability* value is tunable. We have specified the false positive probability as 0.01 (so that the other BF parameters are tuned accordingly). The main reason to implement this phase is to optimize the structure of the input data sets by eliminating unnecessary input splits. Depending on the number of the input split size, the number of map function executors is decided. This phase will be executed between the preprocessing MR job and selection MR job. In general, it is a small process not requiring large computing capabilities. However, if the input data sets become large, then BF execution time increases. To avoid such situation, the parallel version of the semantic BF could be used (as described in Section 3.5.1). In map function, the input data are pruned by removing the unwanted white spaces or removing the dot character from each line. Basically, in this phase the input is split to extract the RDF values from the N-Triple format files. The reduce phase generates BF files corresponding to input files containing the triples. Next, the output of this phase is used as the input for the execution of the BF. In our current implementation this phase can be triggered by using "-fps" (i.e., force preprocessing stage) command.

4.2 Selection phase

Selection phase allows to filter the unnecessary input data and produce output triples that match at least one of the given queries. The output of the selection phase is expressed in the form of key-value pairs. The keys are the query pattern numbers, and the values are the matching triples along with some extra information. This extra information can be exploited to perform the matching with the common join variables. To this end, input triples will be grouped based on the given query patterns, which in turn are represented as BGPs. As soon as the selection phase is completed, the join operation can start (i.e., the output of selection phase will be used as the primary input for the join operation).

Selection phase generates different files for different query patterns. The files are created only when the query pattern selects at least one RDF triple. The newly generated output files are saved on the shared HDFS storage layer. Fig. 8 displays a simple example of a BGP and how selection phase processes two files with names "pattern 1" (i.e., "?author creates ? Novel") and "pattern 2" (i.e., "?author writes ?XYZ") to generate the final output file. These two patterns (which are described by the BGP input) must be matched against the input RDF triples. RDF triples are grouped and split into separated files (here for simplicity only two files are taken into consideration). Both the BGP and the input RDF files are passed to an MR job, which produces an output containing the selected inputs.

4.3 Join phase

During the execution of join phase, RDF graph is generated, based on a given complex query. An RDF graph is a set of RDF triples such a subgraph

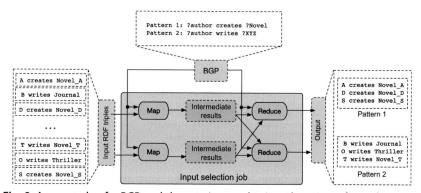

Fig. 8 An example of a BGP and the way input selection phase is performed.

represents a subset of the triples in the main graph. To execute complex join operations, a correct RDF graph must be constructed. After constructing the graph from the given complex query, the graph is read/processed to develop a query execution plan. The join operation is executed based on the generated plan. The joining phase runs iteratively when the BGP consists of joins that have multiple common join variables (subgraphs can be used to perform the join operation more efficiently). Multiway join is an efficient way to execute the complex query. In some cases, it is more efficient to implement a multiway join as a single MR job [39]. In MR, join operations can be implemented as a two steps process: (i) a join plan is generated, and (ii) a dedicated task executes the plan. Query pattern numbers and the common join variables on which these patterns must be joined will be fed into the join process.

4.3.1 Query join plan generation

The query join plan (RDF graph) is generated from the complex queries by taking the query pattern number as the *vertex* and the join variables as the *edge* of the graph. Fig. 9 shows a generated RDF graph from a query. The query is to select all the articles ("?article1") that author ("?author1") has published. The RDF graph is based on two join variables ("?article1" and "?author1") consisting four query patterns (patterns are numbered as mentioned in Fig. 9). As a first step to generate a query plan to process, we will group the query patterns based on common join variables. As we can see query pattern 0, 1, and 2 can be grouped as they share the same single join variable which is "?article1". So we can process this group and called this joining as a first join operation. After joining the query pattern 0, 1, and 2, now we have a new graph called A. This newly generated RDF graph will be reread to

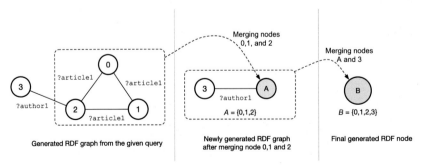

Generated RDF graph from the given query | Newly generated RDF graph after merging node 0,1 and 2 | Final generated RDF node

Fig. 9 Example of the process generating the query execution plan. Note: *Dotted rectangular box* represents the nodes that share common join variable.

execute the next join. From Fig. 9, we can see that query pattern A, and 3 can also be grouped as they have "?author1" as shared join variable. So, we can also perform the join operation based on "?author1" and we will get a final RDF graph called B.

4.3.2 Query join execution

Join operation can be implemented either as a map or a reduce function (see Section 4). However, the reduce-side join can handle a larger amount of data. Multiway join is another form of equi-join. For instance, given two data sets (A and B), the equi-join is defined as a combination of tuple $a \in A$, and $b \in B$ such that $A.a = B.b$, where a and b are respectively the values from columns in A and B on which the join operation is performed. Each join operation is a single MR job divided into map function and reduce function. For the first join operation, the input is the output of the selection task.

Algorithm 1 shows the main steps composing the map phase of the join operation. The mapper receives as input the key-value pairs. The former (i.e., keys) are represented by the query pattern number together with the common join variable; while the latter (i.e., values) are represented by matched triples together with bindings bound to the variable name (i.e., a combination of the bindings and the matched triples). Triples are extracted from the input value if the bindings match with the given join variable. In that case, the map function projects the bindings if they are

---

ALGORITHM 1 Algorithm for the map phase of join operation.

Data: *key* – pattern file name
Data: *value* – list of bindings together with triples
1 Read input value
2 Extract the binding from the value
3 Read the query patterns to join
4 Read the join keys
5 **if** *variable of binding matches with the join variable* **then**
6 | set *key* ← get the value using the variable of binding
7 | set *value* ← get the value using the concatenation of *key* and *value*
8 **end**
9 **return** {*key,value*}

equal to the join key. The variable value (i.e., either the subject, predicate or object position of the triple) will be used as the mapper's output key, and finally, the key and associated values of the map task are sent to reduce phase.

In the reduce phase, the input key is given by a join key value, and the input value is a list of all the values that are associated with the input key. Algorithm 2 shows the algorithm used by the reduce task. The reducer iterates through the list of values for a given input key and fetches the tag. Tags are the keys that are associated with the input triples for the join phase. Tags are used to verify whether the group values come from all the pattern files that are being joined. If a group does not have a value from some pattern files that need to be joined, that group is an invalid group and will be discarded. The reducer outputs the valid group values (i.e., emitted as keys) which are a concatenation of query pattern numbers and the binding together with the triples. After the output is generated, a new MR job is executed to project the final output based on the given condition.

4.4 SELECT operation

The SELECT operation is initiated once the join operation is completed. It is executed to project the required results from the join operation. It also represents the final MR job, allowing to apply the "SELECT" clause

---

ALGORITHM 2 Algorithm for the reduce phase of join operation.

Data: *key* – join key value
    Data: *values* – list of values associated with the input key

1 Get the query pattern numbers to perform join
2 **foreach** *value* ∈ *values* **do**
3 | **set** *tag* ← extract the tag from the input value
4 | Add the tag to the tag set
5 **end**
6 **if** *list of query pattern numbers is equal to the tag set* **then**
7 | Get new table name
8 | Add list of query pattern numbers to the newly created table
9 | **set** *key* ← newly created table
10 | **set** *value* ← get the values contained in the newly created table
11 | **return** {*key,value*}
12 **else**
13 | **return** {*Invalid Group*}

expressed in the query. The map phase of this MR job takes the key–value pairs produced by the join MR job and uses them as input. Here, also binding plays an important role to match the SELECT clause variables. On success, the values of the binding variable become the output values, and the SELECT clause variable becomes the output key. Next, this key–value pair is transferred to the reduce phase to write it back to HDFS.

5. Evaluation

The proposed approach has been tested, and some of the results are shown here to prove its efficacy. We evaluated the proposed framework using last stable version of the Apache Hadoop framework running on a Linux (Ubuntu-based) machine. The host platform was a server equipped with an Intel Core^{TM} i7-6700K processor running at 4.00 GHz. The host platform was also equipped with 32GiB of main memory and 1TiB of storage. This storage space was used to manage the HDFS file system. To assess the performance of the proposed framework, we used the SP^2 benchmark [40]. It is worth to note that for correctness check, the same set of queries were executed on Jena framework (in a standalone system) and they were compared with our results.

5.1 Query patterns

The selected queries are chosen to show that the framework not only works for SELECT but also for DISTINCT and CONSTRUCT operations, which are used by Q1, Q2, Q4, and Q10 use cases.

- Q1 has only one shared join variable among three subqueries and returns the year of publication of "Journal 1 (1940)".
- Q2 is more complex as it has only one shared join key but nine atomic queries. It extracts all inproceedings with nine properties (dc:creator, bench:booktitle, dc:title, swrc:pages, dcterms:partOf, rdfs:seeAlso, foaf:homepage dcterms:issued, and optionally bench:abstract) ordered by the year. This query implements a bushy graph pattern and accesses large strings. The size of the final result grows with the input data sets.
- Q4 has eight atomic subqueries and five common join variables. It selects all distinct pairs of article author names for authors that have published in the same journal. It contains a long graph chain, and the output is huge.
- Q10 is the simplest one with no selection phase. It returns publications and venues in which Paul Erdoes is involved either as author or as editor. Q10 implements an object bound-only access pattern and the result size stabilizes for large data sets.

All the data meant for the SP^2 benchmark are in the N3 format (i.e., Notation 3 format). They are DBLP-like data represented with the N-Triple syntax. The DBLP database contains bibliographic information, and data are formatted as subject, predicate and object. We have used five different data sets that contain respectively 5×10^4, 10^6, 5×10^6, 10^7, and 5×10^7 tuples. All the reported values are an average of three runs.[g] In the following, we describe results obtained considering the mentioned query use cases belonging to the SP^2 benchmark.

5.2 Execution time

The tests are carried out on all the five different data sets, and the reported time values are in seconds. Table 1 shows the performance of the proposed framework when input data is preprocessed using BFs (optimized framework execution), while Table 2 provides the results obtained using our framework but disabling preprocessing with BFs.

Response time in all the executions are different due to the different level of complexity, and it is evident that BF has impacted the results in a positive sense. The space-efficiency exposed by the BF mainly causes this positive effect. The improvements for each query are: Q1 has seen minimum 18% and maximum 55% of improvements over non-BF implementation; similarly Q2 has seen 5% minimum and 37% as maximum, Q4 has seen 8% as a minimum and 22% as a maximum. Finally, the most significant improvement has been seen for Q10 which has seen the improvement from 39% to 82%. It is clear from the results that the filtering phase (i.e., preprocessing input data) has a severe impact on the total execution time for the complex queries. It is worth to note that, the preprocessing stage for all queries takes the same amount of time. However, some of the join and SELECT map-reduce jobs show a similar response time in both the environments. In general, the join phase is the most compute-intensive operation. While the input data sets become large, then also the SELECT operation starts to consume a considerable amount of time, which is due to the multiple query join variables and increasing input data sets (see Table 1).

To further verify the impact of the optimization process using the BF, the most complex queries Q2 and Q4 are picked (see Fig. 10). Here, it is evident that the filtering phase has largely improved the execution time.

---

[g] During the evaluation, the system was not shared by any other tasks.

Table 1 Execution time (in seconds) of each individual jobs using BF

Query	Preprocessing (in seconds)	Selection (in seconds)	Join (in seconds)	SELECT (in seconds)	Total time (in seconds)
Input = 50K triples					
Q1	19	15	17	15	66
Q2	19	24	21	22	86
Q4	19	19	114	22	174
Q10	19	0	12	15	45
Input = 1 million triples					
Q1	48	24	18	16	106
Q2	48	40	25	25	128
Q4	48	30	260	46	384
Q10	48	0	15	18	81
Input = 5 million triples					
Q1	55	36	39	20	150
Q2	55	95	167	73	390
Q4	55	90	480	100	725
Q10	55	0	18	19	92
Input = 10 million triples					
Q1	61	52	43	23	179
Q2	61	138	172	150	521
Q4	61	130	910	195	1296
Q10	61	0	20	20	101
Input = 50 million triples					
Q1	170	203	70	15	458
Q2	170	720	1080	425	2395
Q4	170	680	1703	650	3203
Q10	170	0	33	27	230

Table 2 Total execution time (in seconds) without using BFs.

Query	Q1	Q2	Q4	Q10
Input = 50K triples				
Total time (in seconds)	84	113	207	80
Input = 1 million triples				
Total time (in seconds)	129	165	428	133
Input = 5 million triples				
Total time (in seconds)	192	555	910	299
Input = 10 million triples				
Total time (in seconds)	398	832	1666	570
Input = 50 million triples				
Total time (in seconds)	835	2515	3478	1063

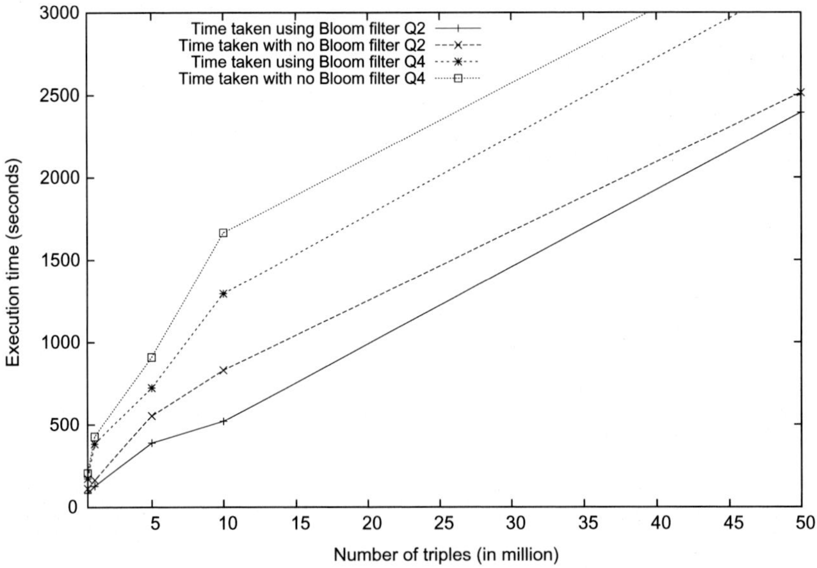

Fig. 10 Comparison of the response times between case using the BFs on Q2 and Q4 queries, and case not using the BF stage.

This scalability has been achieved using efficient algorithms, while the underlying distributed programming model Hadoop still performs well with increased complexity. It is worth to note that only Q1 produces one triple as output, while other queries have produced a large triples as output.

For instance, Q4 reports a huge amount of query size such 2,586,733 triples from one million triples versus Q2 reports 32,770 triples from one million. Similarly, Q4 reports 18,362,955 triples and Q2 reports 248,738 triples from five million.

5.2.1 Application of the framework

The possible application of our proposed model would be to use in Cloud for querying a large amount of RDF data using SPARQL. The current public Cloud-based VMs offers an effective environment (MR ready VMs) for running Hadoop MR applications. Such could be an ideal scenario where the BF part can run inside a single container[h] and the PBF can run using multiple containers. Thus our proposed approach can easily be ported into the public Cloud setup to extract information from complex SPARQL queries.

5.3 Performance improvements using PBF

Aiming at improving the throughput of the proposed query engine, we implemented separately a parallelized version of the Bloom filter (PBF). To assess the performance gain provided by the multithreaded implementation, we compared the execution time for verifying input query memberships for both sequential and multithreaded BFs. To this end, we generated a synthetic data set of 10^5 input queries, while concurrent threads have been implemented using the standard *pthread* library. PBF scalability has been assessed by performing a set of experiments while varying the number of threads from 1 to 128. Since the thread execution is influenced by other processes and threads managed by the scheduler of the operating system and running on the server, we performed three runs for each configuration of the PBF (i.e., running the PBF with a different number of threads). For each run, we collected the execution time of each thread and calculated the mean (μ) and the standard deviation (σ). As expected, increasing the number of concurrent threads that compete for the physical resources increased the execution time of newly launched threads. However, calculating the standard deviation for the pool of launched threads, we obtained a good statistical behavior. In fact, a large fraction of the thread execution times t_{w_i} were in the range $(\mu - \sigma) \leq t_{w_i} \leq (\mu + \sigma)$. For instance, with 128 concurrent threads, we obtain that more than 92% of the thread execution times

---

[h] https://linuxcontainers.org/.

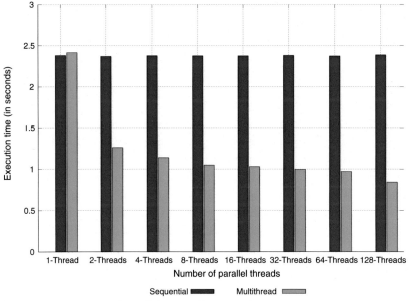

Fig. 11 Execution time of parallel Bloom filter (PBF).

were in the range [0.7102, 1.9854], where $\mu = 1.3478$ and $\sigma = 0.6376$, thus confirming the advantage of the parallel version of the BF.

Fig. 11 shows the comparison of the execution times between the sequential (violet bars) and multithreaded (green bars) BF implementations. The plot clearly shows that multithreaded BF implementation outperforms the sequential one, except for the case with a single thread. This behavior is motivated by the introduction of thread creation overheads that are not present in the sequential version, and that is compensated by the execution time reduction when the number of threads grows. The increased speedup also demonstrates the scalability of the implemented solution (see Fig. 12). Due to the limited number of physical cores on the host server used in the experimental session (i.e., the Intel Core^{TM} i7-6700K CPU is equipped with four hyperthreaded cores, total in eight simultaneous threads), the scalability is less pronounced when the number of concurrent threads is in the range between 16 and 64. Conversely, the parallel BF implementation shows a larger speedup improvement moving from single thread execution to multithread execution using 2 to 8 threads and moving from 64-threads to 128-threads execution. Larger speedups are also expected using server nodes equipped with CPUs exposing more physical cores to the pool of concurrent threads.

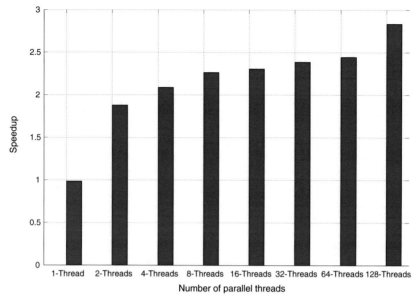

Fig. 12 Speedup of parallel Bloom filter (PBF).

Finally, we also evaluate the scalability of our PBF implementation, by increasing the size of the synthetic input data set. To this end, we performed a set of further experiments varying the data set size from 10^5 to 10^7 entries. The BPF showed a large constant speedup in all the runs, thus confirming the effectiveness of the proposed approach.

Advantages of using sequential BFs have been demonstrated in Section 5.2 with the execution of complex queries Q2 and Q4. The experimental results provided in this section confirm that further performance gain can be obtained by using a PBF implementation concerning sequential one.

6. Conclusion and future work

This chapter proposed a scalable, efficient query engine to execute complex query joins, based on a distributed and fault resilient framework such as the Hadoop Map–Reduce. The proposed engine connects an efficient processing framework, an efficient probabilistic data structure (i.e., Bloom filter), and a simple but efficient joining algorithm driven by the BGP approach. In this chapter, the proposed engine used SPARQL as the query language because of its support for the RDF data format. The SP^2 benchmark has been used to evaluate and show the effectiveness of the proposed approach (the improvement from minimum 5% to maximum 82% running benchmark queries). In this work, we also advocated and showed the efficiency of the BF as a tool for preprocessing input data. To execute complex queries, we have also used the BGP matching concept. In this perspective, it would also be more appropriate to execute disjoint components of the given complex queries in parallel. As future work, we will investigate on varying the false positive probability value of the BFs and comparing its impact on the framework performance. Successfully tuning the false positive value is expected to increase the performance while processing huge data sets largely. Apart from that, this work can be further extended to be used either with Apache Spark jobs or with Mesos by correctly providing the relevant APIs.

Acknowledgments

Part of the work has been done while Somnath Mazumdar was a student at Université Nice Sophia Antipolis, France. The authors thank Fabrice Huet for the initial discussion on the part of the presented idea and useful suggested references. The authors also wish to thank the editor for the helpful and insightful comments, which significantly improved the quality of this chapter.

References

[1] O. Lassila, R.R. Swick, Resource description framework (RDF) model and syntax specification, 1999.
[2] E. Prud, A. Seaborne, et al., SPARQL query language for RDF, 2006.
[3] J. Dean, S. Ghemawat, MapReduce: simplified data processing on large clusters, Commun. ACM 51 (1) (2008) 107–113.
[4] D. Borthakur, J. Gray, J.S. Sarma, K. Muthukkaruppan, N. Spiegelberg, H. Kuang, K. Ranganathan, D. Molkov, A. Menon, S. Rash, et al., Apache Hadoop goes realtime at Facebook, in: Proceedings of the 2011 ACM SIGMOD International Conference on Management of data, ACM, 2011, pp. 1071–1080.
[5] C. Doulkeridis, K. Nørvåg, A survey of large-scale analytical query processing in MapReduce, VLDB J. 23 (3) (2014) 355–380.
[6] H.-C. Yang, A. Dasdan, R.-L. Hsiao, D.S. Parker, Map-reduce-merge: simplified relational data processing on large clusters, in: Proceedings of the 2007 ACM SIGMOD international conference on Management of data, ACM, 2007, pp. 1029–1040.
[7] Z. Kaoudi, I. Manolescu, RDF in the clouds: a survey, VLDB J. 24 (1) (2015) 67–91.
[8] S. Sakr, G. Al-Naymat, Relational processing of RDF queries: a survey, SIGMOD Rec. 38 (4) (2010) 23–28.
[9] L. Yang, L. Yang, J. Niu, Z. Hu, J. Long, M. Zheng, A Semantic Data Parallel Query Method Based on Hadoop, in: W. Cellary, M.F. Mokbel, J. Wang, H. Wang, R. Zhou, Y. Zhang (Eds.), Web Information Systems Engineering - WISE 2016 Springer International Publishing, Cham, 2016, pp. 396–404.
[10] C. Yu, J. Wang, W.-Q. Ling, Design and implementation of a hybrid storage and query system based on Hadoop for massive traffic data, Inf. Technol. Inform. Z1, (2016) 21.

[11] N. Papailiou, I. Konstantinou, D. Tsoumakos, P. Karras, N. Koziris, H2RDF+: High-performance distributed joins over large-scale RDF graphs, in: 2013 IEEE International Conference on Big Data, IEEE, 2013, pp. 255–263.

[12] D. Borthakur, et al., HDFS architecture guide, Hadoop Apache Project 53 (2008) 1–13.

[13] X. Zhang, L. Chen, Y. Tong, M. Wang, EAGRE: Towards scalable I/O efficient SPARQL query evaluation on the cloud, in: 2013 IEEE 29th International Conference on Data Engineering (ICDE), IEEE, 2013, pp. 565–576.

[14] A. Schätzle, M. Przyjaciel-Zablocki, T. Berberich, G. Lausen, S2X: graph-parallel querying of RDF with GraphX, in: VLDB Workshop on Big Graphs Online Querying, Springer, 2015, pp. 155–168.

[15] A. Schätzle, M. Przyjaciel-Zablocki, A. Neu, G. Lausen, Sempala: interactive SPARQL query processing on Hadoop, in: International Semantic Web Conference, Springer, 2014, pp. 164–179.

[16] F. Goasdoué, Z. Kaoudi, I. Manolescu, J. Quiané-Ruiz, S. Zampetakis, CliqueSquare: efficient Hadoop-based RDF query processing, in: BDA'13 - Journées de Bases de Données Avancées, Nantes, France, 2013.

[17] T. Neumann, G. Weikum, Scalable join processing on very large RDF graphs, in: Proceedings of the 2009 ACM SIGMOD International Conference on Management of data, ACM, 2009, pp. 627–640.

[18] B.H. Bloom, Space/time trade-offs in hash coding with allowable errors, Commun. ACM 13 (7) (1970) 422–426.

[19] M. Bermudez-Edo, T. Elsaleh, P. Barnaghi, K. Taylor, IoT-Lite: a lightweight semantic model for the Internet of Things, in: 2016 Intl IEEE Conferences on Ubiquitous Intelligence & Computing, Advanced and Trusted Computing, Scalable Computing and Communications, Cloud and Big Data Computing, Internet of People, and Smart World Congress (UIC/ATC/ScalCom/CBDCom/IoP/SmartWorld), IEEE, 2016, pp. 90–97.

[20] X. Su, H. Zhang, J. Riekki, A. Keränen, J.K. Nurminen, L. Du, Connecting IoT sensors to knowledge-based systems by transforming SenML to RDF, Procedia Comput. Sci. 32 (2014) 215–222.

[21] M.H. Namaki, F. Chowdhury, M.R. Islam, J.R. Doppa, Y. Wu, Learning to speed up query planning in graph databases, arXiv preprint arXiv:1801.06766 (2018).

[22] M. Wylot, M. Hauswirth, P. Cudré-Mauroux, S. Sakr, RDF data storage and query processing schemes: a survey, ACM Comput. Surv. 51 (4) (2018) 84.

[23] M. Husain, J. McGlothlin, M.M. Masud, L. Khan, B.M. Thuraisingham, Heuristics-based query processing for large RDF graphs using cloud computing, IEEE Trans. Knowl. Data Eng. 23 (9) (2011) 1312–1327.

[24] T. Neumann, G. Weikum, RDF-3X: a RISC-style engine for RDF, Proceedings VLDB Endowment 1 (1) (2008) 647–659.

[25] I. Abdelaziz, R. Harbi, Z. Khayyat, P. Kalnis, A survey and experimental comparison of distributed SPARQL engines for very large RDF data, Proceedings VLDB Endowment 10 (13) (2017) 2049–2060.

[26] S. Dolev, P. Florissi, E. Gudes, S. Sharma, I. Singer, A survey on geographically distributed big-data processing using mapreduce, IEEE Trans. Big Data 5, (2017) 60–80.

[27] A. Cuzzocrea, R. Buyya, V. Passanisi, G. Pilato, MapReduce-based algorithms for managing big RDF graphs: state-of-the-art analysis, paradigms, and future directions, in: Proceedings of the 17th IEEE/ACM International Symposium on Cluster, Cloud and Grid Computing, IEEE Press, 2017, pp. 898–905.

[28] H. Han, H. Jung, H. Eom, H.Y. Yeom, Scatter-Gather-Merge: An efficient star-join query processing algorithm for data-parallel frameworks, Cluster Computing 14 (2) (2011) 183–197.

[29] M. Husain, J. McGlothlin, M.M. Masud, L. Khan, B.M. Thuraisingham, Heuristics-based query processing for large RDF graphs using cloud computing, IEEE Trans. Knowl. Data Eng. 23 (9) (2011) 1312–1327.

[30] J. Myung, J. Yeon, S.-G. Lee, SPARQL basic graph pattern processing with iterative MapReduce, in: Proceedings of the 2010 Workshop on Massive Data Analytics on the Cloud, ACM, 2010, p. 6.

[31] A. Schätzle, M. Przyjaciel-Zablocki, S. Skilevic, G. Lausen, S2RDF: RDF querying with SPARQL on spark, Proceedings VLDB Endowment 9 (10) (2016) 804–815.

[32] M. Hammoud, D.A. Rabbou, R. Nouri, S.-M.-R. Beheshti, S. Sakr, DREAM: distributed RDF engine with adaptive query planner and minimal communication, Proceedings VLDB Endowment 8 (6) (2015) 654–665.

[33] Q.-Z. Niu, et al., Optimization of RDF reasoning based on MapReduce, in: World Automation Congress (WAC), 2016, IEEE, 2016, pp. 1–5.

[34] T. White, Hadoop: The Definitive Guide, O'Reilly Media, Inc., 2012.

[35] K. Shvachko, H. Kuang, S. Radia, R. Chansler, The Hadoop distributed file system, in: 2010 IEEE 26th Symposium on Mass Storage Systems and Technologies (MSST), IEEE, 2010, pp. 1–10.

[36] J.J. Carroll, I. Dickinson, C. Dollin, D. Reynolds, A. Seaborne, K. Wilkinson, Jena: implementing the semantic web recommendations, in: Proceedings of the 13th International World Wide Web Conference on Alternate Track Papers & Posters, ACM, 2004, pp. 74–83.

[37] H. Marais, K. Bharat, Supporting cooperative and personal surfing with a desktop assistant, in: Proceedings of the 10th Annual ACM Symposium on User Interface Software and Technology, ACM, 1997, pp. 129–138.

[38] S. Tarkoma, C.E. Rothenberg, E. Lagerspetz, Theory and Practice of Bloom Filters for Distributed Systems, IEEE Commun. Surv. Tutorials 14 (1) (2012) 131–155.

[39] F.N. Afrati, J.D. Ullman, Optimizing multiway joins in a map-reduce environment, IEEE Trans. Knowl. Data Eng. 23 (9) (2011) 1282–1298.

[40] M. Schmidt, T. Hornung, G. Lausen, C. Pinkel, SP^2Bench: a SPARQL performance benchmark, in: IEEE 25th International Conference on Data Engineering, 2009. ICDE'09, IEEE, 2009, pp. 222–233.

About the authors

Somnath Mazumdar received his PhD degree in Computing Systems from the University of Siena, Italy in 2017. He is currently a postdoctoral researcher at Simula Research Laboratory, Norway. His main research interests are heterogeneous HPC computing, computer architectures, performance analysis. He is/was associated with multiple international research projects.

Alberto Scionti received his PhD (European Doctorate degree—2011) and MSc (2007) in Computer and Control Engineering, from Politecnico di Torino. He is co-author of more than 40 publications on books, international journals, conferences, and workshops. Currently, he is a senior researcher at Istituto Superiore Mario Boella (ISMB), focusing on advanced computer architectures, high-performance heterogeneous and low-power systems, distributed and cloud-based computer architectures. His research interests include memory design, reconfigurable computing system architecture, application of evolutionary algorithms to design optimization.

CHAPTER TWO

A study of DVFS methodologies for multicore systems with islanding feature

Shervin Hajiamini[a], Behrooz A. Shirazi[b]
[a]Grinnell College, Grinnell, IA, United States
[b]Washington State University, Pullman, WA, United States

Contents

Advances in Computers, Volume 119
ISSN 0065-2458
https://doi.org/10.1016/bs.adcom.2020.03.005

Abstract

High performance computing centers need to keep up with the growing applications of varying computational characteristics. High computation rates result in consuming vast amounts of energy with increasing electricity costs. To fulfill computational demands with reasonable energy consumption cost, dynamic voltage and frequency scaling (DVFS) technique is used for scaling voltage/frequency (V/F) levels of cores based on their time-varying workloads during application runtime. This book chapter investigates improving multicore system energy efficiency with DVFS, where optimization goal is to minimize application execution time while maintaining the energy consumption below a user-defined energy budget and vice versa. This optimization goal is achieved by performing DVFS at fine-grain level, which adjusts individual cores V/F levels, and at coarse-grain level, which divides the cores into multiple voltage/frequency islands (VFIs), where all the cores in each VFI share a common V/F level. Despite being very energy-efficient, the fine-grain VFIs have high implementation overheads. The coarse-grain VFIs provide acceptable energy efficiency with lower overheads. The fine-grain DVFS establishes energy efficiency optimality used for evaluating the coarse-grain VFIs performance. Factors considered for improving the coarse-grain VFIs energy efficiency include scheduling tasks among VFIs, clustering cores with similar workloads across application execution intervals, and fixing the VFIs V/F levels for the entire application runtime or adjusting them per application execution interval. The fine- and coarse-grain VFIs energy efficiency performances, optimized at compile-time, are evaluated on multiple applications that have varying computational characteristics. This book chapter also evaluates these optimization methodologies scalabilities for systems with different number of cores.

Abbreviations

CG	coarse-grain
CGS	coarse-grain with symmetric
DCG	dynamic coarse-grain
DP	dynamic programming
DVFS	dynamic voltage and frequency scaling
EB	energy budget
EDILP	energy-delay integer linear programming
EDP	energy-delay product
FG	fine-grain
HPC	high performance computing
ILP	high performance computing
MILP	mixed integer linear programming
OS	operating system
ROI	region of interest

SCC   single-chip cloud computer
V/F   voltage/frequency
VDP   Viterbi-based dynamic programming
VFIs   voltage/frequency islands

1. Introduction

This book chapter brings together a comprehensive overview of the focused works that we have conducted and previously published (or to appear) in the literature. These works include [1–4].[a]

The emergence of modern High Performance Computing (HPC) systems with hundreds of multicore chips is driven by increasingly large applications with varying computation during runtime. The growing computational capability of such massive multicore platforms results in a dramatic increase in their energy consumption which, in turn, increases cooling costs and decreases reliability of system components. According to the National Resources Defense Council [5], HPC centers will consume up to 140 billion kilowatt-hours annually by 2020, which costs American businesses $13 billion in electricity bills each year. Reducing the energy consumption is expected to save 39 billion kilo-watt hours, which is equivalent to $3.8 billion per year. However, energy saving comes at the cost of increasing the application execution time. As such, the goal of HPC users often conflicts with that of system managers, as the former is to have their applications executed with minimal turnaround time, while the latter is driven to design a more energy-aware system. Therefore, system designers need to consider the balance between the system energy consumption and the application execution time to meet the system stakeholders' objectives.

The dynamic voltage/frequency scaling (DVFS) is a well-known method that has been used to provide effective balance between the energy consumption and the execution time. As a generic technique, DVFS reduces the voltage/frequency (V/F) level of cores that are idle in some application intervals, e.g., due to inter-core data exchange, to reduce the energy consumption. DVFS increases the V/F levels of cores that have high workloads in other intervals to improve application performance.

---

[a] Some parts of this book chapter, including the text and the figures, are adapted from these prior publications.

Scaling cores V/F levels effectively increases system energy efficiency. The energy efficiency of methodologies discussed in this chapter measures the extent to which the application execution time is reduced such that the system energy consumption does not exceed a user-defined energy budget and vice versa.

Extensive studies have been conducted on applying DVFS on multicore systems. This book chapter focuses on two major aspects of these studies. The first aspect is per-core DVFS, where the V/F levels of cores are adjusted individually. Because of the individual V/F level scaling, this technique is very energy-efficient but may incur high overhead when the cores V/F levels vary frequently. Because this technique fully scales the V/F levels for each individual core at run-time, it can be used to establish a Pareto frontier or optimal base-line for comparing coarse-grain (VFI-based) DVFS methods. The second aspect of DVFS is VFI-based DVFS, where cores are divided into different clusters (islands). In each island, cores share the same V/F level, whereas cores in different islands may operate on different V/F levels. Compared to the per-core DVFS, this architecture has lower runtime overhead and provides reasonable energy efficiency.

For the per-core DVFS, this book chapter investigates two types of methodologies. The first methodology, implemented by integer linear programming (ILP), optimally scales cores V/F levels and achieves the best tradeoff between energy consumption and execution time. However, this methodology is not scalable with increasing number of cores. To speed up DVFS runtime, a heuristic methodology is considered that uses a dynamic programming technique called the Viterbi algorithm. This technique uses overlapping subproblems and memorization concepts for reducing the time of computing cores V/F levels while globally considering cores workloads for balancing energy-time tradeoff. In this chapter, the ILP-based DVFS establishes the best energy-time solutions, presented by a Pareto frontier curve, and is used for evaluating the performance of the Viterbi algorithm and other heuristic algorithms.

There are two major problems to solve in the VFI-based DVFS: (1) clustering cores, and (2) determining the cores V/F levels. This book chapter presents two solutions for solving these problems. The first solution uses the optimized V/F levels for cores workloads per application interval (tasks) that are obtained by the per-core DVFS as explained above. The tasks are then scheduled on cores such that each VFI contains a number of cores that execute tasks with the same (static) V/F level. The second solution dynamically tunes the VFIs V/F levels, similar to DVFS, during the runtime to match the needs of the applications with varying workloads. Furthermore, using the dynamic

VFIs at runtime reduces load imbalance and improves core under-utilization, which may not be addressed by the static VFIs.

This book chapter is organized as follows. Section 2 explains selected important aspects of the energy efficiency optimization for the methodologies introduced in this chapter and related work. Section 3 presents background and assumptions required for design and implementation of DVFS methodologies discussed in the book chapter. Section 4 introduces configuration setups for a simulator and benchmarks used for evaluating the methodologies energy efficiency. Sections 5 and 6 discuss per-core DVFS and VFI-based DVFS methodologies, respectively, which were devised for maximizing the optimization goal defined above. Section 7 concludes this chapter and presents future work.

2. Related work

This section presents an overview of works that devised DVFS methodologies for improving the energy efficiency of multicore systems. These works extensively studied DVFS to gain energy saving in multicore systems with respect to different factors such as the utilization of system components [6], task scheduling and migration [7], compiler optimization [8], thermal profile of cores [9], dynamic resource allocation [10], and virtual core mapping [11]. This section is not a survey report and does not cover all the above mentioned topics. Rather, it performs a high-level comparison to relevant works, considering only factors that are used in this chapter to study the multicore system energy efficiency. Table 1 shows these factors and a brief explanation for each of them.

Table 1 Factors for comparing to the related work.

Factors	Brief explanation
Number of cores	Number of cores in multicore systems
Compile/run-time	Static or dynamic scaling of cores V/F levels
Number of V/F levels	Fine granularity of V/F levels
Optimization parameters	Parameters used for improving the energy efficiency
Energy efficiency metrics	Metrics defined for measuring the energy efficiency
Optimality of energy efficiency	Exact or heuristic methodologies devised for deciding cores V/F levels
Applications	Applications run on multicore systems

Section 1 mentioned that two main aspects of DVFS are discussed in this book chapter: per-core DVFS and VFIs. The remaining part of this section compares our suggested methodologies to the related work for each of the aspects. Each of the sub-sections below explains comparisons, factor-wise, between the suggested and related works.

2.1 Per-core DVFS

Table 2 compares the proposed per-core DVFS methodologies to other works in the literature.

The columns and rows in Table 2 are explained as follows:

2.1.1 Number of cores

The proposed work is evaluated on a system of 64 alpha cores. Spiliopoulos et al. [6] used two system setups: (1) Intel Core i7 that is a quad-core chip, and (2) a quad-core AMD Phenom II. Murray et al. [9] also used 64 alpha cores. Tarplee et al. [12] experimented with Intel and AMD servers that included 4–12 cores. Jung et al. [13] applied their technique to a system of four processing cores. Lai et al. [14] used Intel single-chip cloud computer (SCC) platform that has 48 cores. Chen et al. [15] conducted their experiments on a computer with Intel 8-core CPU. Wang et al. [10] considered a cluster of 3 servers with 6 cores, 6 cores, and 8 cores, respectively,

2.1.2 Compile/run-time

The proposed work performs per-core DVFS at compile-time. Spiliopoulos et al. [6] predicted V/F levels at regular intervals during the application runtimes. To predict V/F levels, they used measures collected from performance counters in the previous interval. They utilized measures to predict energy and execution time of tasks for all V/F levels and selected a suitable V/F level for the next interval. Murray et al. [9] inserted compile-time flags in applications. They used these flags to mark the start of idle periods. When they ran programs, cores V/F levels were not scaled until the flags were activated. Tarplee et al. [12] used multiple compile-time algorithms to generate Pareto frontier curves. Jung et al. [13] devised a run-time strategy for deciding cores V/F levels using a lookup table that maps application task features, e.g., its priority and arrival time, to V/F levels. The lookup table was constructed at compile-time. Lai et al. [14] used a profile-guided strategy to change the cores V/F levels during application runtimes. Wang et al. [10] devised a runtime methodology for distributing application tasks among cores and scaling their V/F levels. Chen et al. [15] proposed a compile-time based technique for scheduling tasks on with suitable V/F levels.

Table 2 List of factors for per-core DVFS methodologies.

Method	Number of cores	Compile-time or run-time	Number of V/F levels	Optimization parameters	Energy efficiency metric	Optimal (O) or Heuristic (H)	Applications
Proposed work	64	Compile-time	10	Energy Time	EDP	ILP/O, Viterbi/H	SPLAH-2 PARSEC
[6]	4	Run-time	9 (Intel) 4 (AMD)	Energy Time	EDP ED^{2}P	Governor/H	SPEC2006
[9]	64	Compile-time and run-time	6	Energy Time	EDP	Heuristic/O	SPLAH-2 PARSEC
[12]	4–12	Compile-time	3	Energy Time	Energy-time tradeoff	ILP/O NSGA–II/O	Synthetic
[13]	4	Compile-time and run-time	4	Energy Time	Energy	Supervised learning/H Value iteration/H	Synthetic
[14]	48	Compile-time and run-time	8	Energy Time Instructions per cycle Bus utilization	EDP	Power manager/H	SPLASH-2 NPB Malstone Synthetic
[10]	6, 8	Run-time	3	Energy Time	Energy-time tradeoff	Markov model/O Falk & Palocsay/H	Synthetic
[15]	8	Compile-time	4	Energy Time	Power	Mixed ILP/O	MiBench E3S Synthetic

2.1.3 Number of V/F levels

The proposed work uses 10 V/F levels from 1.25 to 2.5 GHz. Spiliopoulos et al. [6] used 9 V/F levels from 1.6 to 2.66 GHz for the Intel cores and 4 V/F levels from 0.8 to 3.2 GHz for the AMD cores. Tarplee et al. [12] used 3 V/F levels from 0.32 to 0.46 GHz. Murray et al. [9] utilized 6 V/F levels. Jung et al. [13] applied 4 V/F levels from 0 to 0.25 GHz. Lai et al. [14] used 8 V/F levels from 0.1 to 0.8 GHz. Wang et al. [10] conducted their experiments with 3 V/F levels. Chen et al. [15] utilized 4 V/F levels from 1.01 to 2.1 GHz.

2.1.4 Optimization parameters

The proposed work considers minimizing execution time penalty while maintaining energy consumption below an energy budget. Spiliopoulos et al. [6] maximized energy saving with a limit on execution time penalty. Authors in [9,12,13] optimized energy usage and execution time. Lai et al. [14] considered energy consumption, execution time, bus utilization, and instructions per cycle. Wang et al. [10] maximized profit that was the average response time for serving request from clients subtracted by the amount of energy usage. Chen et al. [15] minimized energy consumption while maintaining the execution constraint of application tasks.

2.1.5 Energy efficiency metrics

The proposed work uses energy-delay product (EDP), where energy the total energy consumption of cores and delay is the amount of time for executing applications. Spiliopoulos et al. [6] used EDP and ED^2P, where execution time was computed as the average execution time over multiple programs that run simultaneously on their multicore systems. Murray et al. [9] used EDP. Tarplee et al. [12] defined Pareto frontiers to tradeoff energy consumption versus execution time. Jung et al. [13] measured energy consumption for different classifiers. Lai et al. [14] computed EDP for each phase of their applications to optimize an EDP target. Wang et al. [10] used profit as explained above. Chen et al. [15] used power consumption.

2.1.6 Optimality of energy efficiency

The proposed work uses integer linear programming (ILP) to obtain optimal solutions and the Viterbi algorithm to obtain heuristic solutions. Spiliopoulos et al. [6] used a heuristic technique, where for each application interval energy consumption and execution time is predicted across all V/F levels. Then, among the V/F levels, the one that minimizes either EDP or ED^2P

is selected for the next interval. Tarplee et al. [12] used linear programming methods and a genetic algorithm for computing optimal solutions. Murray et al. [9] devised a heuristic that scaled cores V/F levels depending on whether the data required for load instruction was fetched from L2 cache or the main memory within a certain number of cycles. Jung et al. [13] proposed value iteration heuristic to build a table that mapped system states to suitable V/F levels. Lai et al. [14] used a heuristic that for each application phase predicted EDP as a function of frequency, instruction per cycle, and bus utilization across V/F levels. Then, this heuristic selected a V/F level that had the most impact on minimizing EDP. Wang et al. [10] proposed a framework that heuristically dispatched tasks among cores and optimally determined cores V/F levels, as well as the number of turned on cores. Chen et al. [15] used a linear programming technique to optimally compute the idle interval of cores for the V/F level scaling.

2.1.7 Applications

The energy efficiency of the proposed work in this chapter is measured on applications that are selected from SPLASH-2 and PARSEC benchmark suites [16,17]. Spiliopoulos et al. [6] used SPEC2006 workloads. Tarplee et al. [12] used non-traditional high performance computing benchmarks that included applications such as file compression, gaming engine, and graphics rendering. Murray et al. [9] used SPLASH-2 and PARSEC applications. Jung et al. [13] executed applications related to TCP/IP. Lai et al. [14] used applications that were selected from SPLASH-2, NAS [18], and Malstone [19] suites, as well as a few synthetic applications. Wang et al. [10] generated applications tasks from a few software applications [15] used MiBench [20], E3S [21], and a few synthetic applications.

2.2 Voltage/frequency islands (VFIs)

Table 3 compares the VFIs methodologies proposed in this chapter to the other works in the literature.

The columns and rows in Table 3 are explained as follows:

2.2.1 Number of VFIs

The proposed work uses both symmetric and asymmetric VFIs for a 16-core system. In the symmetric design, there are four VFIs that each contains four cores. For the asymmetric VFIs, the number of VFIs and their contained cores are determined based on the number of V/F levels from the per-core DVFS optimization or the number of cores that have similar computational

Table 3 List of factors for VFIs methodologies.

Method	Maximum number of VFIs	Compile-time or Run-time	Number of V/F levels	Optimization parameters	Energy efficiency metric	Optimal (O) or Heuristic (H)	Applications
Proposed work	4 (symmetric) Variable (asymmetric)	Compile-time	4 (symmetric) 4, 10 (asymmetric)	Energy Time	EDP	ILP/O Mixed ILP/O	SPLAH-2 PARSEC
[22]	4	Compile-time and run-time	6	Energy Time Traffic	EDP	K-means/H ILP/O Feedback/H	SPLAH-2 PARSEC
[23]	4	Compile-time	6	Energy Time Traffic	EDP	ILP/O Task stealing/H	Phoenix++
[24]	3	Compile-time	4	Energy Time Traffic	Energy-time	Mixed ILP/O Feedback/H	E3S
[25]	12	Compile-time and runtime	5	Energy Time Traffic	Energy-time	Rule-based task scheduling and VFI creation/H	E3S Synthetic
[26]	2	Compile-time	2	Energy Time	Energy-time	Genetic algorithm/H	Multimedia Synthetic
[27]	2	Compile-time	5	Traffic energy and time Floorplanning	Time	ILP/O	MCSL
[28]	1	Compile-time	5	Energy	Energy	Largest task first/H Newton's method/H	Synthetic

behavior across application execution intervals. Kim et al. [22] used four VFIs with variable number of cores per VFI. Duraisamy et al. [23] devised a system with 4 symmetric VFIs, where each VFI contained 16 cores. Ogras et al. [24] experimented with 1 to 3 variable-size VFIs. Jin et al. [25] constructed their VFIs on systems consisting of 16, 36, and 64 cores, which consisted of 4, 6, and 12 VFIs, respectively. Ozen and Tosun [26] portioned their multicore system of up to 16 cores into 2 variable-size VFIs. Demiriz et al. [27] divided their 16- and 64-core system into 2 VFIs, where each VFI contained between 3 and 5 cores. Pagani and Chen [28] used the Intel SCC platform as a single-VFI system.

2.2.2 Compile/run-time
The proposed work suggests compile-time methodologies for creating VFIs and scaling their V/F levels. Kim et al. [22] created VFIs at compile-time and performed DVFS at run-time. Duraisamy et al. [23] defined the V/F levels of VFIs at compile-time. Ogras et al. [24] performed compile-time computations for their VFI-based systems. Jin et al. [25] created initial configuration for VFIs at compile-time, but adjusted VFI sizes and V/F levels at run-time. Authors in [26,27] devised compile-time methodologies. Pagani and Chen [28] created their VFI at compile-time, but changed its V/F level at run-time.

2.2.3 Number of V/F levels
In the proposed work, when VFIs are created based on the per-core DVFS, the V/F levels of VFIs are chosen from 10 V/F levels optimized by the per-core DVFS. When VFIs are created based on the computational similarity among cores, the VFI optimization problem uses 4 V/F levels from 1.25 to 2.5 GHz. The symmetric VFIs use the same 4 V/F levels. Kim et al. [22] considered 6 V/F levels from 1.25 to 2.5 GHz. Duraisamy et al. [23] used the same set of V/F levels. Ogras et al. [24] utilized 4 basic V/F levels from 12.5 to 20 MHz. Jin et al. [25] considered 5 V/F levels from 0.7 to 1.1 V. Ozen and Tosun [26] used 2 V/F levels, 1.8 V and 3 V. Demiriz et al. [27] used 5 V/F levels from 0 to 1 V. Pagani and Chen [28] used 5 V/F levels from 0 to 2 GHz.

2.2.4 Optimization parameters
The proposed work considers energy consumption and execution time. Authors in [22–26] optimized energy consumption, execution time, and inter-core traffic energy and time. Demiriz et al. [27] optimized the

inter-core traffic energy and time, as well as the placement of cores and their types on a chip (floorplanning). Pagani and Chen [28] optimized the energy consumption.

2.2.5 Energy efficiency metrics
The proposed work uses EDP. [22,23] also used EDP. Authors in [24–26] considered energy-time tradeoff. For [27], execution time was their energy efficiency metric. Pagani and Chen [28] used energy consumption.

2.2.6 Optimality of energy efficiency
The proposed work uses an ILP and a mixed ILP for creating VFI and scaling their V/F levels. Kim et al. [22] used K-means clustering [29] for creating VFIs. They set the VFIs V/F levels with ILP and adjusted them using a feedback controller. Duraisamy et al. [23] applied ILP for creating VFIs and scaling their V/F levels. They adjusted the VFIs V/F levels by re-assigning highly utilized cores in a separate VFI. Ogras et al. [24] used a mixed ILP for creating VFIs and performing DVFS. They adjusted the VFIs V/F levels by using a feedback controller. Jin et al. [25] devised a heuristic that created VFIs by assigning tasks to cores with as low V/F levels as possible. Ozen and Tosun [26] used a genetic algorithm for VFI creation and V/F level assignment. Demiriz et al. [27] utilized an ILP. Pagani and Chen [28] applied a heuristic that considered the size of tasks when assigning them to cores in a VFI. They suggested a heuristic for assigning a single V/F level to all of the cores in that VFI.

2.2.7 Applications
The proposed work uses the benchmark suites that were mentioned in Section 2.1. Kim et al. [22] used the same benchmark suites. Duraisamy et al. [23] evaluated their methods on Phoenix++ applications [30]. Ogras et al. [24] chose applications from E3S [21]. Jin et al. [25] utilized E3S [21] and synthetic applications generated by the TGFF tool [31]. Ozen and Tosun [26] conducted their experiments on multimedia [32,33] and synthetic applications. Demiriz et al. [27] used MCSL benchmark suite [34]. Pagani and Chen [28] used synthetic applications.

3. Preliminaries

This section presents the following: (1) An optimality criteria, which is used in this book chapter to find out the degree to which the solutions of optimization frameworks, discussed in this book chapter, are close to optimal solutions, (2) An architecture for the manycore system used for running the experimented applications, (3) An execution model of applications (workloads), (4) A strategy to profile the parameters of applications used by the optimization algorithms to optimize the energy efficiency of VFI-based systems, and (5) A practical use of optimization frameworks for real-world applications.

3.1 Definition of optimality

The problem formulations, which will be discussed in Sections 5 and 6 of this book chapter, provide solutions for performing the per-core and the VFI-based DVFS to minimize the application's execution time (objective) without exceeding the energy budget (constraint). For problem formulations optimizing functions with conflicting trends, i.e., energy vs time, the discussion of optimality is transformed into a Pareto optimality context. The Pareto optimality adopted in this book chapter is defined based on the dominance relationship between any two solutions to a constrained single-objective optimization problem. Fig. 1 shows an example of the dominance relationship among three solutions (A, B, and D) obtained from solving an optimization problem with an objective (o) and a constraint (c), where o is minimized without exceeding c. In Fig. 1, for the same value of c (c_1), A obtains lower value for o (o_1) compared to D (o_2). Thus, A dominates D. On the other hand, neither A nor B dominate the other. It is because the o values of A and B (o_1 and o_3) are obtained under different c values (c_1 and c_2).

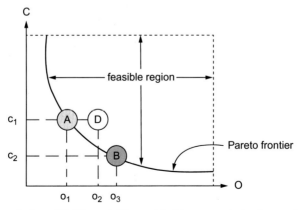

Fig. 1 Three solutions (A), (B), and (D) provided by an optimization problem with an objective (o) and a constraint (c).

Solutions that are not dominated by the other solutions are known as Pareto optimal solutions (i.e., A and B in Fig. 1) and a set of such solutions form a Pareto frontier (dark-colored curve in Fig. 1). The Pareto optimal solutions are of great interest because compared to these solutions, there is not a feasible solution with better objective value unless it is obtained under a different constraint.

Our fine-grain (FG) V/F level assignment problem (see Section 5.2) generates an approximation [12] of such Pareto frontier among all feasible energy-time (makespan) solutions. As it will be explained in Section 5.2, this problem formulation is solved by using integer linear programming (ILP), where the ILP-based Pareto frontier is used as baseline to (1) evaluate the energy-time solutions obtained by a number of per-core DVFS algorithms (Section 5.7), and (2) determine how far/close the energy-time solutions of our coarse-grain VFI systems are compared to comparable solutions obtained from FG Pareto frontier (Section 6.3).

3.2 System model

The multicore system assumed in this book chapter is a set $C = \{c_1...c_N\}$ of cores arranged in an $n \times n$ mesh ($N = n \times n$) that is partitioned into multiple VFIs as shown in Fig. 2. It is assumed that all cores in each VFI operate under a common V/F level that may impact optimizing the overall system performance. Each core has a local non-unified L1 cache and all cores share a unified L2 cache.

Definition 1. The manycore system is partitioned into a set of voltage/frequency islands $I = \{i_1...i_M\}$, where each VFI operates at a V/F level selected from a set VFL $= \{1...L\}$ with VFL $= 1$ and VFL $= L$ denoting minimum and maximum V/F levels, respectively. For the fine-grain configuration,

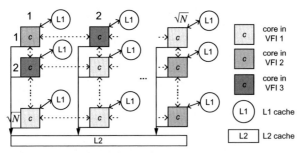

Fig. 2 A multicore system with three islands.

each VFI, i_j, has only one core, i.e., $|i_j| = 1$, whereas for the coarse-grain configuration, a VFI has at least a core, i.e., $|i_j| \geq 1$ and at least one VFI has two or more cores. Thus, VFIs may or may not have the same size in the coarse-grain partitioned manycore system.

3.3 Task execution model

3.3.1 Benchmark's execution characteristics

The applications considered in this book chapter are chosen from benchmarks suites (see Section 4.2) that use multithreaded workloads. Moreover, these benchmarks are designed for distributed memory systems, wherein a shared memory (e.g., L2 cache) is used to facilitate efficient data exchange among the cores (or threads) at runtime [16]. For all execution runs reported in this book chapter, a section inside the benchmark source codes, known as region of interest (ROI), is considered to evaluate our proposed formulations. This region is pre-configured to be executed in parallel by all cores to satisfy the benchmark's concurrency demands.

3.3.2 Task definition

The cores executing the benchmarks have distinct computational phases or execution windows at runtime. At the beginning of each phase/window, dynamic V/F levels (for fine-grain approach) are adjusted to meet the optimization goals. To ensure the correct execution of ROI, the benchmark source codes are instrumented with synchronization routines (such as barriers) to resolve memory access delays and perform data race recovery for the current phase before executing the next one. This book chapter leverages such computational phases, separated by barriers, to define tasks executed on each core, where phase boundaries are established by the barriers. Each phase in this application, which may consist of one or more function invocations, corresponds to a task that represents the core's workload.

Definition 2. An application consists of a set of tasks $T = \{\tau_1 \dots \tau_r\}$, as shown in Fig. 3, where each task τ_i, executed on core c_i ($1 \leq i \leq r$), is composed of a set of subtasks $\tau_i = \{\tau_{i,1} \dots \tau_{i,p}\}$ and whose execution times may include memory access delays for data exchange among the subtasks through the shared memory. Here, r and p denote the number of cores and application phases, respectively. In Fig. 3, during an interval, gray portions show computation periods of cores executing tasks and black portions show the core's idle

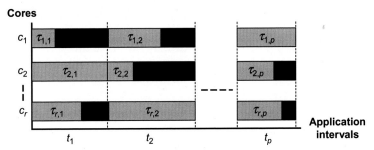

Fig. 3 An application's subtasks with p phases executed on r cores.

periods representing overheads caused by inter-core synchronizations at the end of each interval. The optimization techniques, discussed in this book chapter, take advantage of these variations in idle periods to improve energy efficiency and performance by slowing down cores that execute subtasks with longer idle periods and speeding up cores with subtasks that have shorter idle periods within any given interval.

3.4 Application profiling

The energy-efficiency algorithms studied in this book chapter rely on two application parameters, i.e., subtasks execution times and energy consumption. A subtask execution time during an application phase is the time that a core executes the subtask before reaching a barrier (shaded boxes in Fig. 3.) Energy consumption of a subtask refers to the rate of core power usage during the subtask execution time.

The above two parameters are obtained by profiling the application. The profiling of a benchmark/application collects the execution time and the energy consumption for each application phase in the benchmark for each possible V/F level. It is noted that even though profiling is a time-consuming process (large overhead), it is carried out only once per application and only at compile-time (prior to the actual application execution). Therefore, this overhead is not considered in the complexity of the energy-efficiency algorithms.

3.5 Real-world usage of the energy-efficiency algorithms

Our proposed optimization framework is specifically a good fit for kernel code or embedded applications that can be optimized once at compile-time, and then executed many times over. Therefore, our overall solution steps for all the energy-efficiency algorithms studied in this book chapter consist of the following steps:

1. At compile-time
    i. Profile the application and obtain the execution time and the energy consumption parameters.
    ii. Apply an energy-efficiency algorithm using the profiled parameters to predict the best V/F levels per core/VFI either for the entire application runtime or for each application execution phase.
2. At runtime
    i. Static approach: the V/F levels optimized in Step 1 are assigned to cores/VFIs and stay fixed at runtime.
    ii. Dynamic approach: the optimized V/F levels are stored in a lookup table. At the start of each phase, the system (e.g., operating system (OS) or microkernels) fetches V/F levels from the lookup table within a cycle and issues them with low overhead to voltage regulators that tune the speed of cores in hundredth of a nanosecond.

4. Experimental setup

This section consists of five parts. The first part explains our many-core system configurations used to generate the subtasks execution time and energy consumption profiles. The second part presents benchmarks considered in this book chapter to evaluate the energy efficiency of the per-core/VFI optimization algorithms. The third part explains the configuration set up used to solve the optimization problems. The fourth part defines energy budget that is used as a constraint for the ILP-based problem definitions. The fifth part introduces a metric for evaluating all optimization methodologies discussed in this book chapter.

4.1 Simulation

GEM5 [35] is used to simulate an 8×8 ($N=64$) mesh structure of identical cores whose architectural features are shown in Table 4. The subtask execution times for a V/F level are obtained by running multiple simulations, where in each simulation the cores V/F levels are set to discrete pairs that are selected between 1.25 GH/0.5 V and 2.5 GHz/1.0 V as the lowest and highest V/F pairs in VFL, respectively. It is assumed that V/F level switching time/energy overheads are only about a few hundreds of nano seconds/J order of magnitude [36]. The subtasks energy consumptions are computed

Table 4 Multicore configuration.

Processor	65 nm Alpha cores
L1 cache (instruction and data)	64 Kbytes
L2 cache	8 Mbytes, 128 Kbytes per core
Coherence	MESI
Main memory	512 Mbytes

Table 5 Benchmarks.

Benchmark	Application domain	Problem size
FFT	Fast Fourier transform	1,048,576 data points
RADIX	Integer sort	4,194,304 integers, 1024 radix
WATER	Measure forces and potentials in water molecules	8000 molecules
LU	Dense matrix computation	1024×1024 matrix, 16×16 block
CANNEAL	Routing cost with simulated annealing	200,000 elements

by McPAT [37] given core-level performance data about the benchmark execution phases generated by GEM5.

After the VFIs V/F levels are statically optimized, at runtime, they are fed to a simulated V/F controller that sets the cores' V/F levels at the start of each application execution phase. When our optimized V/F levels are used on a real manycore chip, software drivers (i.e., cpufreq in Linux) control the hardware-based V/F regulators by (re)setting the V/F levels through a special-purpose register interface.

4.2 Benchmarks

The subtasks, defined in Section 3.3, are sampled from a set of benchmarks, shown in Table 5, which for each benchmark provides information about its application domain and problem size. These benchmarks are selected from two benchmark suites, SPLASH-2 and PARSEC, wherein pthread synchronization functions [38] are used to identify the execution phases explained in Section 3.3. In the experimentations shown in this book chapter, the number of subtasks (p) varies from 3 to 40 and the number of cores

(r) is 4, 16, and 64. Thus, the task set size (T) range (see Fig. 3) varies from 24 to 2560 subtasks across these benchmarks. Since each benchmark represents the behavior of a class of programs, multiple runs of similar applications become unnecessary in our evaluation studies.

4.3 Configuration setup for solving the VFIs problem definitions

The fine- and coarse-grain VFI-based problem definitions, which will be explained in Sections 5 and 6, are formulated in AMPL [39] used for modeling and solving large-scale constrained optimization problems. The formulations of these problems are fed to a solver, Gurobi [40], to find a minimum makespan given the execution pattern of applications (see Fig. 3) and the energy budgets constraints.

The coarse-grain VFI definition problem, which will be explained in Section 6.2, is solved by IPOPT [40]. In our experiments, this problem is solved to optimality within a few minutes and the optimal VFIs V/F level assignment solutions are obtained within a few seconds.

4.4 Energy budgets

An optimization problem, used in this chapter, for deciding the V/F levels in the VFI-based systems is to minimize application makespan while maintaining the energy consumption below an energy budget. The energy budget is defined as a reduced energy usage from running benchmarks at the highest V/F level (2.5 GHz/, 1.0 V), i.e., the maximum energy benchmarks consume when running them on a non-DVFS system.

4.5 Energy efficiency metric

To account for the simultaneous impact of energy over time and vice versa, these parameters are combined into a well-known metric, energy-delay product (EDP) [14]. This metric, which measures system energy efficiency, is computed, after obtaining the V/F levels from the per-core and VFI-based optimization methodologies that are discussed in this book chapter.

5. Per-core DVFS

Section 1 discussed that the per-core DVFS methodology dynamically adjusts V/F levels for individual cores in each of the application execution phases. For design and implementation of this methodology, this book chapter takes two perspectives:

(1) Establishing an optimality tradeoff when minimizing both energy consumption and execution time. The optimality tradeoff, which is based the Pareto frontier curve as shown in Fig. 1, is obtained by solving an ILP-based problem.

(2) Proposing a fast dynamic programming (DP) heuristic that overcomes runtime overhead of the ILP-based DVFS when solving the DVFS problem for large applications. Unlike algorithms that make suboptimal V/F level decisions only based on the cores workloads in the current application interval, the proposed heuristic considers cores workloads across all application intervals to perform DVFS.

This section describes methodologies for ILP, DP, and three other heuristics. The performances of these methodologies are compared using the EDP metric explained in Section 4.5.

5.1 Energy-delay integer linear programming

Referring to Section 3.1 that explained the optimality of solutions for solving the single-objective constrained optimization problems, (1)–(3) define an ILP problem for performing per-core DVFS in multicore systems. Section 3.4 explained that the energy consumption and execution time profiles are used by the optimization methodologies discussed in this book chapter. Therefore, the formulation presented below is called energy-delay ILP (EDILP):

$$\text{Minimize } E_i = \sum_{j=1}^{n} \sum_{l=1}^{L} e_l\left(t_{i,j}\right) \cdot x_l\left(t_{i,j}\right) \ \forall i \tag{1}$$

Subject to:

$$D_i = \sum_{j=1}^{n} \sum_{l=1}^{L} d_l\left(t_{i,j}\right) \cdot x_l\left(t_{i,j}\right) \leq \delta \ \forall i \tag{2}$$

$$\sum_{l=1}^{L} x_l\left(t_{i,j}\right) = 1 \ \forall t_{i,j} \tag{3}$$

where, L and T denote total number of V/F levels and time intervals, respectively. It is assumed that $l=1$ and $l=L$ represent the lowest and highest V/F levels, respectively. In (1) and (2) $e_l(t)$ and $d_l(t)$ are respectively profiled energy consumption and execution time of the core under V/F level l at time interval t. $x_l(t)$, a solution of EDILP, is a binary variable, which determines whether the V/F level l is assigned to the core at the time interval t. The inequality (2) indicates that the application execution time, D_i, is constrained by an upper-bound time δ. The upper-bound time is an additional time relative to the application time at the highest V/F level L. Eq. (3) indicates that at most 1 V/F level is assigned to the core at each interval. Given optimal V/F levels from solving (1)–(3), the total energy consumption and execution time of the multicore system is obtained as follows:

$$E = \sum_{i=1}^{m} E_i \tag{4}$$

$$D = \max_i (D_i) \tag{5}$$

where, E and D are summed energy consumption and maximum execution time over the number of cores, C, respectively. The pseudocode of EDILP is shown in Algorithm 1. The input for this algorithm is organized as two matrices, e and d (Line 1), whose rows represent C cores and columns correspond to T time intervals over which the energy consumptions/execution times are profiled for $|VFL|$ V/F levels.

Given these inputs, the energy consumptions and execution times for each core c_i is extracted from the above mentioned matrices, $e(i, 1)$ to $e(i, L \times T)$ and $d(i, 1)$ to $d(i, L \times T)$, are passed to the ILP solver, which determines whether V/F level l is assigned to a task executed on core i at time interval j, denoted by a binary/decision variables $x_l(t_{i,j})$ (Lines 4–5). Having obtained decision variables values, the core energy consumption and execution time are computed (Line 6). Meanwhile, the algorithm keeps track of cumulative energy consumption and maximum execution time (E and D in (4) and (5)) of the multicore system (Lines 7–9), which are returned as algorithm outputs after the V/F levels of all cores are optimized by the solver.

5.2 A dynamic programming heuristic

This section describes the DP algorithm that uses a global optimization to solve fine-grain (per-core) DVFS problem (1)–(3). This DP algorithm, called Viterbi [29], solves large sub-problems in the next stage. At the final stage, the Viterbi-based DP (VDP) moves backward, through the sup-problems solutions, to obtain the best solution for each of the stages.

Algorithm 1. Pseudocode of EDILP

---

1. **Input.** e and d
2. **Input.** δ
3. **Output.** E and D
4. **for** each core c in C **do**
5. $x_l(t_{i,j}) \leftarrow$ LP_solver ($[e(i, 1), e(i, L \times T)], [d(i, 1), d(i, L \times T)]), \delta),$
   $\forall l \in$ VFL, $\forall t_{i,j} \in T$
6. Compute core energy usage and execution time, E_i and D_i, as shown in (1) and (2).
7. $E \leftarrow E + E_i$
8. **if** $D_i > D$ **then**
9. $D \leftarrow D_i$
10. **end if**
11. **end for**
12. **return** E and D

This section describes VDP for finding the best per-core V/F levels over applications execution phases at runtime. The proposed technique uses EDP as its cost function to determine energy consumption and execution time tradeoff at each application execution phase. The VDP algorithm has polynomial time complexity and low implementation overhead, particularly when the number of V/F levels is less than the number of application execution phases.

VDP operates on a trellis diagram basis (Fig. 4), spanned over P steps (intervals), where in each step j system state s_j is in one of the VFL states.

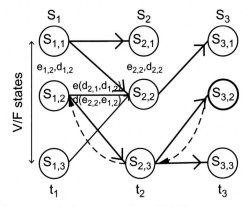

Fig. 4 The trellis diagram with three intervals and 3 V/F states.

A state is defined per-core and per-application execution phase. The state is a pair of energy consumption and execution time measured over multiple V/F levels. A high state suggests that at a high V/F level a core consumes more energy to execute its task while executing the task faster, whereas a low state indicates that at a low V/F level the core consumes less energy to executes its task but it pays for delaying task completion.

Viterbi operates on the trellis in two phases: forward and backward. In the forward phase, in every step j and for every state l, the objective function value (e.g., EDP in our study) is computed for all state sequences starting from a state in the first step and ending in the l-th state at the j-th step where the corresponding objective value is stored. After reaching the last step (P), Viterbi uses these locally stored objective values to backtrack the trellis and find a state sequence, among the other sequences, that provides the best global solution.

Given core state s_{j-1} in current interval $j-1$ and possible states s_j in the next interval j, the cost (C) of the path that consists of a state sequence starting from s_1 and leading to s_j through s_{j-1} is computed as follows:

$$
\begin{aligned}
C\left(s_{1,k} \rightarrow s_{j-1,l'}, s_{j,l}\right) = & f^n\left(E_{j-1,l'}, \Delta\left(e_{j-1,l'}, e_{j,l}\right), e_{j,l}\right) \\
& f^m\left(D_{j-1,l'}, \Delta\left(d_{j-1,l'}, d_{j,l}\right), d_{j,l}\right) \quad \forall l.l' \in \text{VFL}
\end{aligned} \tag{6}
$$

where, symbol $\rightarrow$ indicates the state sequence from $s_{1,k}$ to $s_{j-1,\,l'}$ on the trellis. $E_{j-1,\,l'}$ and $D_{j-1,\,l'}$ are cumulative energy consumption and execution time (corresponding to the state sequence) up to state $s_{j-1,\,l'}$ at V/F level l'. Symbol Δ shows instantaneous transition energy and time costs between states $s_{j-1,\,l'}$ and $s_{j,1}$ at V/F levels l' and l, respectively. $f(.)$ is used to add up and normalize the energy and time variables mentioned above. Integer parameters m and n weigh the energy consumption to execution time portions of the cost function (6).

Having computed the path costs that cross through all the states $s_{j-1,\,l'}$ and $s_{j,l}$ in intervals $j-1$ and j, respectively, the state $s_{j-1,\,l'}$ on the path with the minimum cost is recorded for each state $s_{j,l}$:

$$
s_{j-1,\,l'}^{best} = \underset{s_{j-1,l'}}{\arg\min} \left\{ C\left(s_{1,k} \rightarrow s_{j-1,l'}, s_{j,l}\right) \right\} \quad \forall s_{j,l} \in \text{VFL} \tag{7}
$$

where, $s_{j-1,\,l'}^{best}$ is the state, among all states $s_{j-1,l}$ (corresponding to V/F levels l'), that minimizes the cost function of a state sequence up to interval j. As shown in Fig. 4, among all paths that reach V/F state $s_{2,2}$ in t_2, the one that comes from V/F state $s_{1,1}$ in t_1 (bold arrow) has the lowest EDP compared to $s_{1,2}$ and

$s_{1,3}$. Therefore, $s_{1,1}$ in t_1 is recorded as the backpointer of $s_{2,2}$ in t_2. Of note, to simplify Fig. 4, only the bold arrows are depicted for the other V/F states. The best state sequence (V/F levels) is obtained after determining state s_P, in the P-th interval (last interval), with minimum cost and moving backward on the trellis:

$$s^{best}_{j-1,l} = b\left(s^{best}_{j,l}\right) \qquad (8)$$

where, $s^{best}_{j,l}$ is the best state (V/F level) in interval j. In Fig. 4, V/F states $s_{1,2}, s_{2,3}$, and $s_{3,2}$ (connected by dotted arrows) constitute the best state sequence for t_1, t_2 and t_3. $b(.)$ is the backtracking function, which saves previously visited states $s^{best}_{j-1,l}$ with cost function values (6) being the minimum among the other states on a path from t_1 to t_{j-1}.

5.3 Greedy

As a variation of VDP, the policy of greedy algorithm is based on considering the current states of cores to make future state predictions (local optimality). This algorithm uses the same state definition as VDP to evaluate the energy and time tradeoff over all the states. The general functionality of greedy algorithm is similar to VDP except that the final V/F state sequence prediction is performed based on (6) and (7) rather than the backtracking through the trellis (8). Furthermore, only one state $s_{j-1,l'}$, with the lowest cumulative energy and time ($E_{j-1,l'}$ and $D_{j-1,l'}$ in (6)), is used to predict the next state, $s_{j,l}$. Therefore, the final predicted state sequence is formed by appending a new predicted state, in each interval, to one state sequence, which has the minimum cumulative energy consumption and execution time up to the current interval.

5.4 Feedback controller

In this methodology, in general, prediction error is incorporated in the objective function to prevent unnecessary V/F level changes caused by short-term workload variations at runtime. For the feedback controller, instead of defining a single state for each core based on the energy and time of the subtask executed by that specific core in an interval, a state consists of EDPs that are measured over all the intervals and under multiple V/F levels. For each core, for a given interval, EDPs are computed from all V/F levels. Next, the average of EDPs is computed. Then, the average EDPs are clustered using the K-means algorithm [29], where each cluster defines a state that has a range of EDPs. These clusters are stored in a lookup table, where the lowest and highest V/F levels are mapped to the clusters with average lowest and highest EDPs, respectively. This mapping maximizes energy saving by slowing down under-utilized cores that decreases cores EDPs and improves performance by speeding up highly loaded cores that increases EDPs. We deploy the well-known exponential weighted moving average algorithm [41], which implements the feedback controller to predict core EDP in the next interval based on the core EDP in the current interval while accounting for the prediction error. Once the core EDP for the next interval is predicted, the lookup table that associates core EDPs to V/F levels is consulted to determine the V/F level for the next interval.

5.5 History-based algorithm

This algorithm predicts for each core its EDP for the next interval based on the average of core EDPs for the current and previous intervals. To account for the variability of the cores EDPs across the intervals, the standard deviation of EDPs is incorporated when predicting the next V/F levels. The same lookup table used for the feedback controller is also used here to convert predicted EDP levels to V/F levels.

5.6 VDP time and space complexity

As mentioned in Section 5.2, the VDP algorithm consists of two phases, where in the first phase (forward phase), shown in (6), the costs of state sequences on the trellis are computed and in the second phase (backward phase), shown in (8), the trellis is backtracked to produce optimal state sequence. The backward phase takes constant time because the back-pointers are used to track the states with optimum cost in each interval. The VDP algorithm forward phase is used to track the states with optimum cost in each interval. The VDP algorithm forward phase computes the EDP, $|\text{VFL}|^2$ times in every interval. Thus, for N intervals the time complexity is $O\left(|\text{VFL}|^2 \times N\right)$. Since determining minimal-cost path requires $|\text{VFL}|^2$ comparison pairs and $|\text{VFL}|$ comparison outcomes, $b(s_j)$, are stored for states in every interval, the VDP space complexity is $O\left(|\text{VFL}|^2 + |\text{VFL}| \times N\right)$.

5.7 Performance evaluation

This section compares the performance of the per-core DVFS methodologies described in Sections 5.2–5.5 for multiple applications (see Section 4.2). These methodologies are compared using EDP as the performance metric.

To compute EDP, for each application, energy consumption and execution time are obtained using (4) and (5) based on the V/F levels that are optimized by the methodologies.

Before discussing performance results, the following states configuration parameters of the methodologies used in our experimentations:

For the VDP and greedy algorithms, we use $E^n \times D^m$ as objective function for optimizing energy and time parameters. Thus, the n and m integer factors are used to assign more weight to either energy or time, respectively and indicate the importance of one parameter over the other. In our studies, we would either assign 1 or 2 to m and n because the effect of larger values quickly diminishes beyond the factor of 2.

The feedback controller predicts the core EDP based on a weighted sum of EDP in the current interval, as well as core EDP prediction history. The weighting specifies the amount of contribution of the EDP prediction history over the past intervals relative to the core EDP in the current interval to predict the core EDP for the next interval. Ratios between weights corresponding to the prediction history and the current interval are 0.25, 1, and 4.

The history algorithm predicts the core EDP for the next interval based on 1, 2, or 3 past intervals. This algorithm also uses the lookup table, constructed for the feedback controller, to predict the core EDPs.

Section 3.1 discussed the optimality of the energy-time solutions using the Pareto frontier curve. For an optimization problem with two optimization parameters, e.g., time (c) and energy (o) in Fig. 1, the dominance relation between solutions A and D specifies that A dominates D because for one of the parameters, e.g., c, A provides a value not worse than what D provides (c_1) while for the other parameter, o, A provides a better/less value (o_1) compared to the value provided by D (o_2).

The ILP-based DVFS provides dominant (optimal) solutions in the energy-time search space as discussed in Section 5.2. Referring to Fig. 1 and the previous paragraph, this section demonstrates the dominance of an ILP solution over the other methodologies solutions by (i) presuming that ILP provides as bad execution time delays as the other methodologies, but (ii) quantifying the degree to which ILP solutions have lower-energy values compared to those of VDP, greedy, feedback controller, and history algorithms.

For solving the DVFS problem, each of the heuristic methodologies may come up with a different energy-time solution. As such, each methodology is compared to a corresponding ILP solution on the frontier curve as shown in Fig. 5. In Fig. 5, for each heuristic, the black marker represents an average

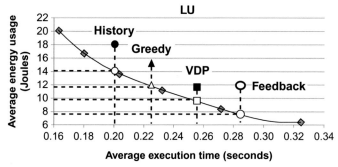

Fig. 5 Computing optimal solutions on the LU Pareto frontier, which correspond to the algorithms solutions.

energy-time solution, where the average energy or time is computed over all configuration parameters as explained above. For an algorithm, e.g., VDP, its corresponding optimal (dominant) solution, provided by EDILP, is a point on the Pareto frontier curve that has the same execution time delay as VDP but has the lowest energy consumption (square white marker in Fig. 5).

Fig. 6 shows the algorithms energy efficiencies, where for each white box, whose length is the EDP of an EDILP solution on the frontier curve, a gray box computes a ratio between the EDP of a dominated heuristic solution and the EDP of the dominating EDILP solution. Needless to say, EDILP obtains the best energy efficiency among the algorithms across all applications. Fig. 6 shows that on the average VDP gains the closest performance to its optimal solution ($1.09 \times$ compared to optimal) among the other algorithms. This indicates that for the same time penalty as the optimal solution, VDP performs DVFS with the best energy saving. The reason is that VDP performs global searches by performing forward and backward passes over the trellis (Fig. 4) while the other heuristic algorithms perform local searches that lead to partial, sub-optimal solutions.

Among all the applications, for WATER and LU, the history and the feedback controller show the worst performances compared to the other algorithms. This is because in WATER and LU, the cores have high workload variations across the intervals, which contribute to the misprediction of the future workloads and the V/F levels. In contrast, for CANNEAL and RADIX, the EDPs of history and feedback are better than or close to VDP. This is due to the consistent workloads of the cores across the intervals, which reduce the error of predicting the future workloads and the corresponding V/F levels.

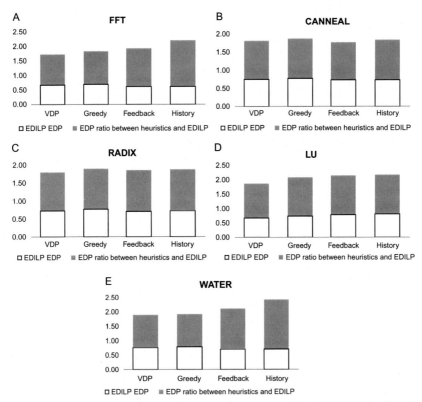

Fig. 6 The EDILP EDPs and ratios between heuristics and EDILP EDPs for (A) FFT, (B) CANNEAL, (C) RADIX, (D) LU, and (E) WATER.

6. VFI-based DVFS

Despite the high energy efficiency obtained by the ILP-based methodologies, they have noticeable runtime overhead when used to optimize large applications. Furthermore, the heuristic algorithms are not practical to optimize the energy efficiency for large systems. The reason is that at hardware-level, each core V/F level is adjusted by a designated regulator that consumes energy, depending on its size, and incurs a delay for V/F level switching. The larger the system is, the more overhead the regulators add to actual energy consumption and execution time.

Section 1 explained that VFIs provide an alternative approach for systems with dynamically adjusted V/F levels. A VFI-based system consists of multiple islands that each contains a number of cores sharing the same V/F level that is controlled by one regulator. VFIs may not be as energy-efficient as the fine-grain DVFS systems, but they provide more economic system design that reduces the overheads mentioned above.

Optimizing the V/F levels of VFIs can be accomplished either at design/manufacturing time (static) or during the application runtime (dynamic). In statically tuned VFIs, each island runs at a fixed V/F level during the entire execution. In dynamically tuned VFIs, similar to DVFS, the island V/F level is adaptively tuned during the runtime to match the needs of the applications with varying workloads. Although the static VFIs provide less expensive design compared to the dynamic VFIs, they are practical for running applications with less dynamic workload variations.

This section describes optimizing the V/F levels for systems with VFI granularity. The static VFIs are defined by scheduling and running tasks on cores/VFIs with fixed V/F levels. The dynamic VFIs are defined by fine-tuning the V/F levels of VFIs that each contains cores with similar execution workloads. This section also compares scalability and performance of these VFI-based systems.

6.1 Static VFI-based systems

This section presents two methodologies for designing the static VFIs. The first methodology is built on the fine-grain V/F levels obtained by EDILP, where applications subtasks with pre-optimized V/F levels are scheduled on the cores and define VFIs. Because of less flexibility of VFIs on adapting individual cores V/F levels, fast execution of complex applications in the high performance computing environments becomes more challenging. As such, the remaining of this book chapter re-defines the optimization goal of EDILP as to optimize cores V/F levels to minimize the application execution time (makespan) for a given energy budget (see Section 4.5). The second methodology schedules the subtasks on pre-configured, symmetric VFIs to achieve the above optimization goal.

To determine VFIs and optimize their V/F levels using the above methodologies, it is assumed that in the execution model (see Fig. 3) the application subtasks that are executed on each core have dependency relations.

6.1.1 VFIs definition with task scheduling and pre-optimized V/F levels

As mentioned above, the first design uses subtasks energy consumptions and execution times obtained by the ILP methodology. Furthermore, it is assumed that after scheduling subtasks on cores, all cores whose assigned subtasks have the same V/F level are placed in one VFI. In a sense, the number of VFIs is determined by the number of unique V/F levels obtained by EDILP. For this VFI design, an optimization problem is defined as: given the subtasks V/F levels and their dependency relations, schedule them on cores such that the makespan is minimized. Because the start time of the scheduled subtasks is non-integer, the above problem is formulated by a mixed integer linear programming (MILP).

The application makespan is determined by the slowest core. The MILP problem constraints that each core executes the subtasks with the same pre-optimized V/F levels. The dependency relations among the subtasks are preserved whether the subtasks are scheduled on the same core or different cores.

6.1.2 VFIs definition with task scheduling and a priori islands V/F levels

In a multicore chips manufacturing process, system designers may opt to use pre-configured islands, where the number of cores in each island and the associated V/F level is fixed during the production phase of the chips. Commercial multicore chips layouts are typically shaped either as square or rectangular because of simpler inter-VFI data transfers among the cores [24,42].

This section assumes a square-shape VFI-based system wherein all VFIs, each one associated with a unique V/F level, have an equal number of cores (symmetric VFIs). For this version of VFIs, an MILP problem is defined as: given symmetric islands, their V/F levels, and the subtasks dependency relations, schedule the subtasks on VFIs such that the makespan is minimized while the energy consumption is maintained below the energy budget.

6.2 Dynamic VFIs

Section 6 explained that the static VFIs have poor performance for the applications with varying workloads. Furthermore, performing subtasks scheduling to define VFIs complicates solving the VFI-based optimization problems. In addition, for large applications, the simultaneous subtasks scheduling and V/F level optimization increase solutions times.

This section addresses the above limitations by determining VFIs over the cores with similar execution workloads across the application intervals. The V/F level of each VFI is then dynamically adjusted during the application execution. The consistency of cores workloads per-VFI reduces energy and speeds up workload execution at the same rate for the contained cores depending on application phase computation demands. This book chapter determines VFIs and performs DVFS in separate steps to reduce the complexity of solving the overall problem.

The optimization problem for determining VFIs is defined by the following nonlinear programming formulation:

$$\text{Minimize} \sum_{j=1}^{P} \sum_{i=1}^{N} \sum_{k=1}^{M} \left(\frac{\tau_{k,j} - t_{i,j}}{\tau_{k,j}} \right) \cdot x_{i,k} \tag{9}$$

$$t_{i,j} \cdot x_{i,k} \leq \tau_{k,j} \quad \begin{matrix} 1 \leq j \leq P \\ \forall c_i \in C_i \\ \forall i_k \in I \end{matrix} \tag{10}$$

$$\sum_{i=1}^{N} x_{i,k} \geq \theta \;\; \forall i_k \in I \tag{11}$$

$$\sum_{k=1}^{M} x_{i,k} = 1 \;\; \forall c_i \in C \tag{12}$$

where, $t_{i,j}$ is the workload (subtask execution time) of core c_i in execution phase j. $\tau_{k,j}$ refers to a core with the maximum workload, at execution phase j, among all cores placed in VFI i_k. $x_{i,k}$ determines whether core c_i is placed in VFI i_k. θ denotes the minimum number of cores per-VFI.

The optimization goal (9) determines VFIs, where each VFI consists of cores whose workloads, across all application phases, are as close as to the highest loaded core in that VFI that is identified in (10). As shown in (9), the workload similarity among the cores per-VFI is determined by minimizing a fraction of time (a nonlinear term) that each core stays idle (after completing the execution phase) relative to the workload of the slowest core in that VFI. To prevent wasting hardware resources for under-loaded VFIs, (11) specifies a minimum number of cores per-VFI.

The following formulation dynamically optimizes the VFIs V/F levels:

$$\text{Minimize} \;\; ms = \sum_{j=1}^{P} \tau_j \tag{13}$$

$$\sum_{l=1}^{L} t_{k,l,j} \cdot x_{k,l,j} \leq \tau_j \quad \begin{matrix} 1 \leq j \leq P \\ \forall i_k \in I \end{matrix} \tag{14}$$

$$\sum_{j=1}^{P} \sum_{k=1}^{M} \sum_{l=1}^{L} e_{k,l,j} \cdot x_{k,l,j} \leq EB \tag{15}$$

where, *ms* is the application makespan. τ_j is the execution time of the slowest VFI in execution phase *j*. $t_{k,l,j}$ and $e_{k,l,j}$ denote the execution time of VFI i_k, which is based on the slowest core in that VFI, and the energy consumption under V/F level *l* at execution phase *j*, respectively. $x_{k,l,j}$ determines whether V/F level *l* is assigned to VFI i_k at execution phase *j*. *EB* denotes the energy budget.

The optimization goal (13) minimizes the application makespan, which is computed as the cumulative execution times of the slowest VFIs across the execution phases. Constraint (14) determines the slowest VFI in each execution phase. Constraint (15) maintains the total energy consumption below the energy budget.

6.3 Performance evaluation

This section evaluates the performance of the VFI-based systems discussed in the previous sections. It also demonstrates how far away the VFIs performances are from the optimal performance of the per-core DVFS. To optimize the V/F levels of the static and the dynamic VFIs, the energy budget (EB) is set to 22.5% energy reduction from the applications maximum energy usages at the highest V/F level.

In the legend of the Fig. 7 that illustrate the performance evaluations, the static VFIs discussed in Section 6.1.1 is referred to as coarse-grain (CG), the static VFIs discussed in Section 6.1.2 as coarse-grain with symmetric VFIs (CGS), and the dynamic VFIs discussed in Section 6.2 as dynamic coarse-grain (DCG). The red lines shown in Fig. 7 indicate the best performances obtained by the per-core DVFS methodology (EDILP) for each of five applications.

Fig. 7A compares the EDPs of CG, CGs, and DCG relative to the per-core DVFS for all applications. Of note, the per-core DVFS is a special case

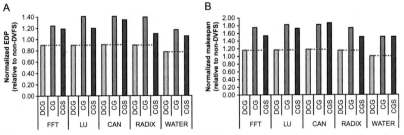

Fig. 7 Comparing (A) energy efficiency and (B) makespan of DCG versus CG and CGS with reference to the per-core DVFS (EDILP).

of DCG, where each core is placed in a separate VFI whose V/F level is dynamically adjusted at runtime. Fig. 7A suggests that the DCG performance is comparable to the per-core DVFS, while on average the CG and CGS performances are more than 1.3 × worse than DCG. Fig. 7B shows the makespans of the applications, running on the static and the dynamic VFIs. Fig. 7B indicates that the DCG performance is on average within 1% of the per-core DVFS across all applications, while the CG and CGS makespans are on average 1.5 × longer than DCG.

The DCG substantial performance gain, compared to the CG and CGS performances, is due to the dynamic, rather than static, V/F level tuning of VFIs during the applications runtimes. Furthermore, DCG runs the application subtasks according to the execution model (see Fig. 3), where the current application phase subtasks are executed only after completing the previous phase subtasks. This execution model has less runtime overhead compared to executing application subtasks that have dependency relations as explained in Sections 6.1.1 and 6.1.2.

7. Conclusion

This chapter summarized state-of-the-art works and suggested several methodologies for optimizing the tradeoff between the energy consumption and the execution time in the multicore systems. The suggested methodologies improve system energy efficiency by minimizing the application makespan under energy usage constraint and vice versa. To achieve this goal, these methodologies optimally perform DVFS on the fine-grain individual cores with high overhead and implementation complexities or optimize the V/F levels for the coarse-grain islands of cores with less design complexity. For the fine-grain DVFS, the results suggest that among all the heuristic techniques investigated in this book chapter, dynamic programming obtains the closest performance compared to the Pareto optimal solutions for the energy-time tradeoffs. The results demonstrate that the Viterbi algorithm, which uses dynamic programming for the V/F level predictions, is particularly useful for applications with sufficient computational variations. For the coarse-grain DVFS, the results suggest that incorporating the task scheduling to solve the islanding and the V/F level assignment problems creates a substantial overhead on the application makespans. Furthermore, the results show that creating VFIs based on the per-time-interval computational similarities, and performing DVFS on VFIs, obtains a performance comparable to the per-core DVFS. Based on the studies discussed in this book chapter, future work will address the following topics:

(1) Improving the energy efficiency of the coarse-grain systems by creating VFIs where the cores in each VFI have consistently similar computations across the time intervals. This can be achieved by sorting the tasks, per-time interval, based on their computational needs, and then assigning the sorted tasks to the cores.

(2) Studying the impact on systems energy efficiency when various performance parameters for the core workload are chosen. A few examples of performance parameters representing a core workload are core utilizations, execution times, and the number of instructions per cycle.

(3) Improving the system energy efficiency by analyzing the degree to which the computational characteristics of the applications benefit from the static or dynamic V/F level optimization of VFIs.

References

[1] S. Hajiamini, B. Shirazi, A. Crandall, H. Ghasemzadeh, C. Cain, Impact of cache voltage scaling on energy-time Pareto frontier in multicore systems, Sust. Compu. Info. Syst. J. (2017) 54–65.

[2] S. Hajiamini, B. Shirazi, A. Crandall, H. Ghasemzadeh, A dynamic programming framework for DVFS-based energy-efficiency in multicore systems, IEEE Trans. Sust. Comput. (2019) 1–12.

[3] S. Hajiamini, B. Shirazi, H. Dong, and C. Cain, "Optimality of dynamic voltage/frequency scaling in many-core systems with voltage-frequency islands," Sust. Compu. Info. Syst. J. 2019, 1–20.

[4] S. Hajiamini, B. Shirazi, C. Cain, H. Dong, An energy-constrained makespan optimization framework in fine- to coarse-grain partitioned multicore systems, in: IEEE Proc. International Green and Sustainable Computing Conference, 2017, pp. 1–8.

[5] P. Delforge, America's Data Centers Consuming and Wasting Growing Amounts of Energy (online), 2015, Available: https://www.nrdc.org/resources/americas-data-centers-consuming-and-wasting-growing-amounts-energy (accessed 1.10.2016).

[6] V. Spiliopoulos, S. Kaxiras, G. Keramidas, Green governors: a framework for continuously adaptive DVFS, in: IEEE Proc. Green Computing Conference and Workshops (IGCC), 2011, pp. 1–8.

[7] R.I. Davis, A. Burns, A survey of hard real-time scheduling for multiprocessor systems, ACM Comput. Surv. 43 (2011) 1–44.

[8] D. Mosse, H. Aydin, B. Childers, R. Melhem, Compiler-assisted dynamic power-aware scheduling for real-time applications, in: Workshop on Compiler and Operating Systems for Low Power, 2000, pp. 1–9.

[9] J. Murray, T. Lu, P. Wettin, P.P. Pande, B. Shirazi, Dual-level DVFS-enabled millimeter-wave wireless NoC architectures, J. Emerg. Technol. Comput. Syst. 10 (2014) 1–27.

[10] Y. Wang, S. Chen, H. Goudarzi, M. Pedram, Resource allocation and consolidation in a multi-core server cluster using a Markov decision process model, in: IEEE Proc. International Symposium on Quality Electronic Design (ISQED), 2013, pp. 635–642.

[11] C. Bae, L. Xia, P. Dinda, J. Lange, Dynamic adaptive virtual core mapping to improve power, in: Proc. International Symposium on High-Performance Parallel and Distributed Computing, 2012, pp. 247–258.

[12] K. Tarplee, R. Friese, A.A. Maciejewski, H.J. Siegel, E.K.P. Chong, Energy and Makespan tradeoffs in Heteregenous computing systems using efficient linear programming techniques, IEEE Trans. Parallel Distrib. Syst. 27 (2016) 1633–1646.

[13] H. Jung, M. Pedram, Supervised learning based power management for multicore processors, IEEE Trans. Comput. Aided Des. Integr. Circuits Syst. 29 (9) (2010) 1395–1408.

[14] Z. Lai, K.T. Lam, C.L. Wang, J. Su, PoweRock: power modeling and flexible dynamic power management for many-core architectures, IEEE Syst. J. 11 (2) (June 2017) 600–612.

[15] G. Chen, et al., Energy optimization for real-time multiprocessor system-on chip with optimal DVFS and DPM combination, in: ACM Transactions on Embedded Computing Systems, 2013, pp. 1–21.

[16] C. Bienia, S. Kumar, J.P. Singh, K. Li, The PARSEC benchmark suite: characterization and architectural implications, in: Proceedings of the 17th International Conference on Parallel Architectures and Compilation Techniques, 2008, pp. 72–81.

[17] S. Woo, et al., The SPLASH-2 programs: characterization and methodological considerations, in: ISCA, 1995, pp. 24–36.

[18] D.H. Bailey, et al., The NAS parallel benchmarks—summary and preliminary results, in: Proc. ACM/IEEE Conf. Supercomputing (SC'91), 1991, pp. 158–165.

[19] C. Bennett, R.L. Grossman, D. Locke, J. Seidman, S. Vejcik, Malstone: towards a benchmark for analytics on large data clouds, in: Proc. 16th ACM SIGKDD Int. Conf. Knowledge Discovery and Data Mining, 2010, pp. 145–152.

[20] M.R. Guthaus, J.S. Ringenberg, D. Ernst, T.M. Austin, T. Mudge, R.B. Brown, MiBench: a free, commercially representative embedded benchmark suite, in: Proceedings of the 2001 IEEE International Workshop on Workload Characterization (WWC), 2001.

[21] K.S. Vallerio, N.K. Jha, Task graph extraction for embedded system synthesis, in: Proceedings of the 16th International Conference on VLSI Design (VLSID), 2003.

[22] R.G. Kim, W. Choi, Z. Chen, P.P. Pande, D. Marculescu, R. Marculescu, Wireless NoC and dynamic VFI codesign: energy efficiency without performance penalty, IEEE Trans. Very Large Scale Integr. VLSI Syst. 24 (2016) 2488–2501.

[23] K. Duraisamy, et al., Energy efficient MapReduce with VFI-enabled multicore platforms, in: ACM Proc. Design Automation Conference, 2015, pp. 1–6.

[24] U.Y. Ogras, R. Marculescu, D. Marculescu, E.G. Jung, Design and management of voltage-frequency island partitioned networks-on-chip, IEEE Trans. Very Large Scale Integr. VLSI Syst. 17 (2009) 330–341.

[25] S. Jin, S. Pei, Y. Han, H. Li, A cost-effective energy optimization framework of multicore SoCs based on dynamically reconfigurable voltage-frequency islands, ACM Trans. Des. Autom. Electron. Syst. 21 (2016) 1–14.

[26] M. Ozen, S. Tosun, Genetic algorithm based NoC design with voltage/frequency islands, IEEE Proc. AICT (2011) 1–5.

[27] A. Demiriz, N. Bagherzadeh, O. Ozturk, Voltage island based heterogeneous NoC design through constraint programming, Comput. Electr. Eng. 40 (2014) 307–316.

[28] S. Pagani, J.J. Chen, Energy efficient task partitioning based on the single frequency approximation scheme, in: IEEE Real-Time Systems Symposium, 2013, pp. 308–318.

[29] C.M. Bishop, Pattern Recognition and Machine Learning, Springer, 2006.

[30] J. Talbot, et al., Phoenix++: modular MapReduce for shared-memory systems, in: Proc. of the Second International Workshop on MapReduce and Its Applications, 2011.

[31] TGFF, http://ziyang.eecs.umich.edu/dickrp/tgff/.

[32] M. Janidarmian, A. Khademzadeh, M. Tavanpour, Onyx: a new heuristic bandwidth-constrained mapping of cores onto tile network on chip, IEICE Electron. Express 6 (1) (2009) 1.

[33] K.-C. Chang, T.-F. Chen, Low-power algorithm for automatic topology generation for application-specific networks on chips, Comput. Digit. Tech. 2 (3) (2008) 239.

[34] W. Liu, J. Xu, X. Wu, Y. Ye, X. Wang, W. Zhang, et al., A noc traffic suite based on real applications, in: IEEE Computer Society Annual Symposium on VLSI (ISVLSI), 2011, pp. 66–71.

[35] N. Binkert, et al., The gem5 simulator, SIGARCH Comput. Archit. News 39 (2011) 1–7.

[36] W. Kim, M.S. Gupta, G.Y. Wei, D. Brooks, System level analysis of fast, per-core DVFS using on-chip switching regulators, in: IEEE International Symposium on High Performance Computer Architecture, 2008, pp. 123–134.

[37] S. Li, et al., McPAT: an integrated power, area, and timing modeling framework for multicore and manycore architectures, in: Presented at the 42nd Annual IEEE/ACM International Symposium on Microarchitecture, 2009, pp. 469–480.

[38] C. Bienia, Benchmarking Modern Multiprocessors, (PhD dissertation), Dept. CS, Princeton Univ., Princeton, NJ, 2011, pp. 1–153.

[39] D.M. Gay, The AMPL modeling language: an aid to formulating and solving optimization problems, in: Proceedings in Mathematics & Statistics, vol. 134, Springer, 2015, pp. 1–22.

[40] AMPL Optimization. (2018). Solvers (online). Available: https://ampl.com/products/solvers/

[41] itl.nist.gov, EWMA Control Charts (online), 2012 Available: http://www.itl.nist.gov/div898/handbook/pmc/section3/pmc324.htm (accessed 20 August 2017).

[42] J. Howard, et al., A 48-Core IA-32 processor in 45 nm CMOS using on-die message-passing and DVFS for performance and power scaling, IEEE J. Solid-State Circ. 46 (2011) 173–183.

About the authors

Shervin Hajiamini is an assistant professor of Computer Science at Grinnell College. His research interest is applying optimization techniques and heuristics for increasing the energy efficiency of multicore systems. He is a member of IEEE.

Behrooz A. Shirazi serves as the NSF CISE/CNS Program Director for the Industry University Collaborative Research Centers Program. With a rotator appointment at NSF, he is also a professor of Computer Science at Washington State University (WSU). He served as the director of the School of Electrical Engineering and Computer Science at WSU from 2005 to 2016. Prior to joining WSU, he served as the Chair of the Computer Science and Engineering Department at University of Texas at Arlington. Dr. Shirazi has conducted research in the areas of pervasive computing, health analytics, green/energy efficient computing, and high-performance computing over the recent years. He is currently serving as the Editor-in-Chief for Sustainable Computing: Informatics and Systems journal. He is the principal founder of the IEEE Symposium on Parallel and Distributed Processing (later joined with IPPS to form IPDPS); a co-founder of the IEEE International Conference on Pervasive Computing and Communications (PerCom); and a co-founder of the International Green and Sustainable Computing Conference (IGSC).

CHAPTER THREE

Effectiveness of state-of-the-art dynamic analysis techniques in identifying diverse Android malware and future enhancements

Jyoti Gajrani[a], Vijay Laxmi[a], Meenakshi Tripathi[a], Manoj Singh Gaur[b], Akka Zemmari[c], Mohamed Mosbah[c], Mauro Conti[d]

[a]Malaviya National Institute of Technology, Jaipur, India
[b]Indian Institute of Technology, Jammu, India
[c]LaBRI, Bordeaux INP, University of Bordeaux, Bordeaux, France
[d]University of Padua, Padua, Italy

Contents

Advances in Computers, Volume 119
ISSN 0065-2458
https://doi.org/10.1016/bs.adcom.2020.03.002

Abstract

Since its launch in 2007, Google's open source mobile operating system Android has become the most prominent OS for smartphones. Availability of 3 million Android apps on official repository, Google Play Store, and a not too tightly controlled environment for app developers have added to the popularity of Android and growth of Android devices. This, however, has also provided an opportunity for malware writers to create inroads into Android devices through malicious apps on App stores including Google Play. These malicious apps may access and leak sensitive information such as details of calls, SMS, emails, pictures, contacts, location, password, etc. Loss of this personal data may lead to fraud, financial loss, threatening, etc. Various solutions based on static, dynamic, or hybrid analysis are proposed by state-of-the-art in the last decade. However, malware writers have also come up with ingenious ways of circumventing detection tools. Recent malware deploy threats like obfuscated and encrypted code, dynamic code loading, and reflection, etc. which fail static analysis approaches employing bytecode for analysis. Dynamic analysis is robust against these evasive methods because it executes the application in the controlled environment. In this chapter, we review dynamic analysis techniques for Android and evaluate these experimentally. We discuss various antidetection methods used by recent Android malware to circumvent even dynamic analysis. We compare the effectiveness of various state-of-the-art dynamic analysis techniques against antidetection techniques. With this chapter, we try to highlight issues and challenges concerned to Android malware analysis techniques that require the attention of research community to avoid loss of end user.

1. Introduction

Nowadays, "Smartphones" have become the most significant part of our lives. Among various mobile technologies, Android leads with 87.7% market share in Q2 2017 [1]. In 2010, commutative app download from Google PlayStore (Android's official online store) was just a billion. But that number had reached a whopping 65 billion in 2016 [2]. However, these figures have magnetized the rapid development of various malicious apps for Android platform. Android's unconstrained nature of installation permits installation of apps downloaded from various third–party app stores. The malware authors make use of this feature and repackage popular apps with malicious codes using reverse engineering. These repackaged apps are then uploaded to various third–party stores. The malicious activities involve frauds, financial losses, criminal activities, privacy leaks, resource drainage, stalking, kidnapping, and many other criminal offenses. Android is coming up with various security patches in its new versions and the new versions are released frequently. In spite of this, most of the users are still using old Android versions.

1.1 Motivation

Among various malicious threats, privacy leakage is the most widely used threat. It stands at a high percentage of 62% [3]. The research community has developed numerous techniques for detection of malware causing privacy leaks. These techniques are based on either static analysis, dynamic analysis, or a combination of both (hybrid analysis). Static analysis techniques read and parse app bytecode for analysis of malware without executing the code. In contrast to this, dynamic analysis runs the app in a controlled environment to monitor its behavior. In this chapter, we mainly focus on dynamic analysis-based systems. The motivation behind considering primarily dynamic analysis originated from three main observations:

1. Advance antidetection methods such as encryption, reflection, obfuscation, and self-code modification are fused by malware to sidestep static analysis. FakeInstaller, a real-world malware uses highly obfuscated code to send expensive premium rate SMS [4]. The elimination of semantic information makes it infeasible for static analysis tools to analyze such apps. According to the report of Virus Bulletin [5], 27% of malware apps use encryption (out of 460,493 analyzed). Andrubis [6] reported that 57.08% of Android malware samples (analyzed in 4 years) employ reflection. Recently, Hammad et al. [7] performed large-scale empirical study on the effects of code obfuscations on Android Apps. The authors analyzed 73,362 obfuscated apps using 61 antimalware products using static analysis. Findings show that code obfuscation significantly impacts Android antimalware products and the majority of antimalware products fail with trivial obfuscations.

2. Existing surveys have given less attention to the assessment of literature based on latest malware tactics such as native code analysis, antiemulation, dynamic code loading, self-modifying code, and automated testing. Specifically, automated testing lies at the core of any dynamic analysis technique due to huge number of apps on different stores. Significant work for automated analysis has been accomplished for Android in recent years. However, we could find only one survey, namely AndroTest [8] which focused on automated testing for Android.

   There are quite a good number of studies available which have discussed DDoS attacks, but there is no specific survey available to consider and gather solutions specific to utility computing models like cloud.

3. The reviews for Android dynamic analysis techniques such as Neuner et al. [9], AndroTest [8] have investigated research work prior to 2014. Since then, sophisticated malware appeared and necessitated reexamining detection techniques.

1.2 Contributions

This chapter attempts to review and evaluate the research work specifically in dynamic analysis focusing:
- Dynamic and hybrid techniques for Android malware analysis (up to 2018).
- Antidetection approaches used by recent malware.
- Comparison of most compelling dynamic analysis systems with respect of various features such as scalability, antiemulation, overhead, etc.
- Empirical evaluation of effectiveness of dynamic sandboxes (which are available online or offline) against mentioned antidetection approaches.
- Open research issues requiring attention of research community.

1.3 Organization

The article is organized as follows: Section 2 briefly discusses Android basics. Section 3 portrays review of existing surveys related to Android malware analysis. We present antidetection approaches used by recent malware in Section 4. In Section 5, we systematically present various dynamic analysis techniques. Section 6 delineates taxonomy of dynamic analysis systems. Substantial number of articles have been studied while writing this chapter. The section includes important contributions in each of the category of taxonomy. We present the evaluation of these systems for antidetection techniques in Section 7. We discuss open research issues of the field in Section 8. Finally, we conclude the work in Section 9.

2. Background

Android has released 30 versions (API levels) by February 2020, since its first release in 2008. The new versions include improvements in form of performance, bug fixing, security updates, GUI enhancements, hardware and software features. The common thing across all versions is the structure of Android application (app). This section briefly explains Android app structure, execution, and communication.

2.1 Android app structure

The four main components of Android app are—(1) Activity is the user interface (UI) that is visible to the user, (2) Service is the component that runs in the background, (3) Content Provider facilitates storage of data, (4) Broadcast Receivers for receiving notification of system events like SMS received, battery low, and events by other apps. Android security model defines a list of permissions for accessing various device resources from components. A resource access is granted to an application only if the user accepts the corresponding permission. Besides predefined permissions, Android facilitates custom permissions using `permission` tag. A developer can protect any component with custom permission which restricts the access by other apps not having the permission as shown in Listing 1. The content provider is protected with custom permission `CP_CUSTOM_PERM` having protection level `dangerous`. This means that any other component or app which want to use component *NewContentProvider* has to explicitly request for granting permission. Besides *dangerous*, Android provides three more protection levels. The *normal* protection level is default and granted without any explicit request. The *signature* protection level permits only apps from same developer to access the component and the *signatureOr System* protection level is special case of signature level permitting access to system apps also.

An apk (installable zip file of application) consists of four essential files and folders as shown in Fig. 1. The first is `AndroidManifest.xml` file that consists of information about all components and permissions required by the application. The name of launcher activity (first activity loaded on app start)

Listing 1 Custom permission in AndroidManifest file.

```
1 <permission android:name= "CP_CUSTOM_PERM" android:
    protectionLevel= "dangerous"/>
2 <provider android:name= "NewContentProvider" android:
    enabled= "true" android:exported= "true" android:
    permission= "CP_CUSTOM_PERM"/>
```

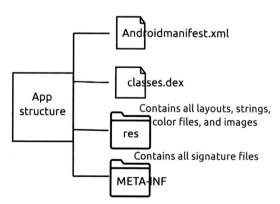

Fig. 1 Android application structure.

is also specified in it. Second, `classes.dex` file containing compiled app code is executed by Dalvik Virtual Machine (DVM). DVM is Android's VM that is better optimized than Java Virtual Machine (JVM) in terms of memory and battery life. Dalvik compiles `dex` file into native instructions when user interacts with app. The third is the `res` folder that consists of all information of UI, layout, images, etc. and the fourth is `META-INF` folder that consists of all signature-related information of app.

2.2 App execution and communication

In Android, each application executes within its own virtual machine (sandbox) as shown in Fig. 2. Sandbox environment is based on privilege separation model and provides app isolation. Starting from API level 21, Dalvik is replaced by Android Runtime (ART) [10]. ART uses ahead-of-time (AOT) compilation to compile entire application into native machine code during its installation instead of on-demand compilation of DVM. This compiled native code is only executed during app interaction. This saves battery life and improves responsiveness due to elusion of unnecessary compilations every time the app interaction occurs. Android performs three main tasks after the app is launched by user. The first is the creation of new process for the app. The second is loading all necessary classes required by application in memory. The last is sending message to launcher activity of app. The app execution starts from the `onCreate()` method of launcher activity and finishes with `onDestroy()` method.

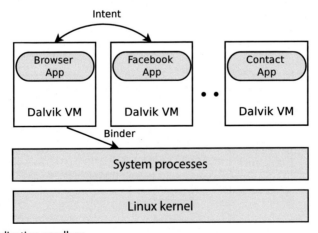

Fig. 2 Application sandbox.

Two or more components (or apps) communicate through Intent, an Inter-Component Communication (ICC) feature provided by Android [11]. Intent is higher level mechanism (user level) that is used by developers in applications. At the level of kernel, OS uses Binder [12] for implementing Intent. Binder is a low level mechanism used by OS to facilitate application communication. A component may call target component by explicitly specifying the name of the components class using an Explicit Intent. Instead of explicitly selecting target with its name, a general action is written using an Implicit Intent. The Android system selects those target components that have specified the matching action in their declaration in the manifest file. The actions are specified for components in manifest file through intent-filters as shown in Listing 2. The Android system selects those target components that have specified the matching action in the declaration in the manifest file. There are some predefined actions such as android.intent.action.BOOT_COMPLETED, android.intent.action. DEVICE_STORAGE_LOW, which can be sent by the system apps only. All the apps that have defined intent-filter for action android.intent.action. DEVICE_STORAGE_LOW, receive a signal whenever the Android system sends it.

Listing 2 Intent-filter declaration in AndroidManifest file.

```
1 <intent-filter>
2    <action android:name= "com.example.service"/>
3 </intent-filter>
4 <intent-filter>
5    <action android:name= "android.intent.action.
      BATTERY_LOW"/>
6 </intent-filter>
```

3. Review of existing surveys

This section presents the review of existing surveys specifically related to Android malware analysis and distinguishes with our contribution. An early survey by Enck [13] reviews various techniques focused on the security analysis of apps for Android. This survey, however, refers to research articles up to 2011. However, considering that many research articles have been published since 2011, this survey can no longer be considered state of the art.

Neuner et al. [9] compared 16 dynamic sandboxes based on implementation details (Android Version), analysis type (static, tainting, GUI Interactions), and analyzed features (file, network, phone, native code). The authors evaluated eight online dynamic sandboxes against eight malware samples and checked whether these sandboxes could identify these malware samples. However, the survey did not evaluate various antidetection techniques that we are focusing in this chapter. In addition to this, we provide empirical evaluation of various offline tools having successful installation.

A survey by Darell et al. [14] presented a taxonomy of contributions based on app deployment stages, i.e., app development time, app availability on the stores, app installation time, app execution, and app update time. The survey is limited to research proposals up to 2014 and lacks evaluation for latest malware tricks. Faruki el al. [15] presented mechanisms used by Android for security enforcement, threats to the these security enforcements, malware growth between 2010 and 2014, and various stealth techniques employed by the malware. Our article presents a comprehensive state-of-the-art survey that supplements as well as complements this review by emphasizing different antidetection techniques incorporated by recent malware to obstructs their analysis. We experimentally evaluate the effectiveness of state-of-the-art dynamic or hybrid analysis techniques in analyzing such malware.

AndroTest [8] provides a comprehensive summary, comparison of various Android test generation techniques up to 2014. The paper evaluates six tools according to four criteria, i.e., code coverage, fault detection capabilities, ease of use, and compatibility with multiple Android versions. The paper concludes that Monkey tool [16], which is provided along with Android by default, is the winner amongst the compared test input generation tools. This is because of Monkey's best coverage (on average), reporting of largest number of failures, ease of use, and compatibility. The paper mentions that other tools need proper implementation and can be combined with the Monkey to achieve significant improvements in code coverage. Our chapter supplements it by including recent work based on advanced intelligent testing strategies up to 2018. The most recent survey by Tam et al. [17] present taxonomy and description of Android malware analysis techniques as static, dynamic, and hybrid analysis techniques.

Table 1 Highlights of existing surveys.

Work	Year	Focus	Highlights
Enck [13]	2011	Platform Security	Discussion of general smartphone (Android, iOS) security issues. Security features provided by smartphone OS. Brief discussion of static, dynamic, permission, and cloud-based application analysis approaches.
Neuner et al. [9]	2014	Dynamic Analysis	Comparison of 16 sandboxes based on level of introspection, functionality, and interdependency.
Darell et al. [14]	2015	Analysis Frameworks	Survey of state of the art up to 2014. Taxonomy based on App development stages. Highlight limitations of various analysis frameworks.
Faruki el al. [15]	2015	Coverage of various issues	Good time-line of major malware families arise during 2010–2013. Tool comparison based on methodology, goal, deployment, and availability.
AndroTest [8]	2015	Automated Testing	Empirical evaluation of test input generation frameworks up to 2014 using 60 real-world applications. Identified strengths and weaknesses of these frameworks.
Tam et al. [17]	2017	Theoretical Evaluation	Theoretical analysis of security solutions published between 2011–2015 based on methodology, scalability, data-set, and sturdiness.

Authors present theoretical analysis of security solutions published between 2011 and 2015. The survey focuses on malware tactics used for complicating static analysis. To complement the work, we have added malware tactics for complicating dynamic analysis, more recent research proposals, and empirical evaluation in addition to theoretical. Table 1 summarizes the highlights of existing surveys.

4. Antidetection methods

Android security solutions based on dynamic analysis approaches can be effective against encryption, obfuscation, memory exploitation, etc. However, recent malware incorporate more advance antidetection techniques to evade detection by dynamic analysis systems. This section discusses various antidetection techniques used by Android malware.

- *Native code*: Android permits the use of native code written in C/C++/ Shell along with Java code. Both native code and Java code allocated in the same address space with the DVM and has direct read and write access to the entire process memory. However, the execution of native code occurs outside the DVM, and malicious activities implemented using native code are not analyzed by systems that work at Dalvik level. Malware from CarrierIQ, DroidPak, and ChathookPtrace families incorporate malicious actions like keystroke logging, location tracking, apk loading, malicious code injection through native code [18]. According to Check Point Software Technologies report [19], Android.Lotoor is in the top three "Most Wanted" mobile malware in the second quarter of 2017. Android.Lotoor exploits vulnerabilities in Android to gain root privilege access on compromised Android devices. The whole malicious code is part of shell scripts [20] that can be identified if the technique can handle native code. Recent work by Mengtao et al. [21] shows that 86% of top 50 Google Play Store apps native code. The other study identified that 16.46% of Google Play apps (out of 227,911 downloaded) use native libraries [22]. The percentage is, even more, i.e., 24% in Asian third–party apps [23]. Therefore, merely the presence of native code is not the mark of maliciousness. Analyzing the native code itself is necessary.

- *Antiemulation*: Due to constraints of battery, storage, processing power, and restoration time in real devices, emulators are preferred for dynamic analysis. The downside is that the device-specific features like model, SDK (Software Development Kit), manufacturer, IMEI (International Mobile Equipment Identity), etc. can be queried by malware to identify emulated environment [24]. This prevents the execution of malicious activities in the emulated environment. *Pincer* is example of Android malware family exhibiting antiemulation behavior [25]. Jing et al. proposed automated heuristics to detect emulated environment [26]. A robust dynamic analysis system must have the methods to handle emulator detection approaches.

Hu et al. came up with two methods—(1) Android source code modification and (2) run-time hooking for handling antiemulation [27]. The former changes variables and APIs behaviors in Android source code and builds the modified source code to generate system's image file, i.e., `system.img`. Android emulator is loaded with this system image.

However, this solution has several limitations like time required for downloading, debugging, and building entire source code. Also, different Android versions need different source code modifications. The latter approach, i.e., run-time hooking dynamically modifies API's behavior by hooking call to APIs [25] and returning the values resembling real devices. The approach is lightweight, flexible, easy to develop, debug, and deploy. Research community proposed Xposed framework [28], Cydia Substrate framework [29], and adbi framework [30] for implementing run-time hooking.

Recent survey on evasion techniques against automated malware dynamic analysis by Alexei et al. [31] concludes that evasive malware is still a major challenge for academic and industry researchers. The survey emphasized to work on the malware that use advance antiemulation techniques.

- *Deep-hiding*: Android apps contain rich and complex GUI widgets. Malware may hide malicious activity deeper in the code or may program to launch at the event associated with some complex widget. Dynamic analysis tool must thoroughly explore GUI to ensure that all reachable codes get executed albeit in multiple executions. The simplest method is to use manual exploration, but it restricts scalability of analysis. The MonkeyRunner tool [32] provides APIs in Python which can be programmed to control GUI exploration from outside the Android code, but it is primarily designed for unit test cases. Monkey [16] is provided with Android mainly for large-scale testing as it generates the pseudo-random sequence of events. But its limitation is that it repeatedly produces the same type of events and cannot work for complex Android GUI elements such as login, text-inputs, scrolls, etc. The research community has proposed various systematic and intelligent exploration techniques which are discussed in Section 6.2.
- *Self-code modification*: Self-code modification is the process of altering program's own instructions during execution [33]. Malware is found to replace benign method call with malicious method call at run-time. This is normally performed in native code language, i.e., assembly language. Droidbench [34] includes malicious samples using self-code modification.
- *Logic bombs*: Malware writers can design code that delivers malicious payload only at various triggers such as time-period, a hard-coded pattern in received SMS, GPS location, etc. Static analysis may detect the presence

of triggers and further guide dynamic analysis. TriggerScope [35] is the first step toward detecting logic bombs for Android using static analysis. TriggerScope analyzes conditional statements/predicates that guard malicious behavior instead of analyzing malicious behavior itself. TriggerScope classifies triggers as suspicious or benign based on the semantic analysis of predicates. However, obfuscation of predicates limits the approach.

- *Compatibility*: The implementation of various approaches proposed in the literature is done by modifying the Android framework. However, a single released version may not work for other API levels if its implementation varies from one API level to next. Approaches having compatibility with all Android versions is a challenge for the success of dynamic analysis.
- *Covert channels*: Two or more applications may employ covert channels to hide the communication used for escalating the privilege of individual app. Marforio et al. [36] implemented various covert channel based attacks like UNIX socket discovery, threads enumeration, processor frequency, etc. It is confirmed by authors that current techniques do not provide a complete defense against different covert channels. Analysis of leaks through these channels is an open problem for the research community.
- *Control flows*: These are more challenging to capture as compared to data-flows. As shown in Listing 3, there is no direct dependency between variables *devId* (contains sensitive data) and *sendId* (sensitive data which is leaked further). However, the value of variable *devId* is being propagated to variable *sendId* by using first binary conversion and then predicates of control flow.

Listing 3 Leak through control flow.

```
1 String devId = tM.getDeviceId();
2 long LdevId = Long.parseLong(devId);
3 String BdevId = Long.toBinaryString(LdevId);
4 String sendId= "";
5 for(i=BdevId.length()-1; i>=0; i--){
6   if (BdevId.charAt(i) ==  '1') {
7     sendId.append( '1'); }
8   else {
9     sendId.append( '0');}
10 }
```

- *Collusion*: Malware hinder the analysis techniques based on single app analysis by spliting the malicious activity across two or more apps. These apps collude at runtime to steal or leak the sensitive data. Analysis techniques targeting one app at a time will not detect such maliciousness. In such cases, all colluding apps must be analyzed simultaneously as an individual app is not malicious. Collusion can be through Intent or can be through covert channels.

- *Reflection*: Reflection is a feature of Java language and can be used in Android apps for instantiating classes, invoking methods/constructors of class, or accessing fields of the class at runtime. Reflection accepts these as parameters of Reflection APIs where parameters can be constructed dynamically or can be encrypted by malware [37]. Research community proposed approaches [4, 38, 39] for analysis of such malware, however multilevel reflection can still fail these approaches.

- *Repackaging libraries*: Most of the Android developers leverage third-party libraries for speedy development and providing advanced features. Intelligent malware seek for vulnerable libraries and exploit such libraries to propagate malicious code. Some malware authors create malicious libraries which when used in benign apps, misuse apps' permissions and perform various malicious activities. Some malware writers also create fake libraries which seems similar to benign libraries. FakeUM is a malware that falsifies it as famous library named UMeng [40].

5. Dynamic analysis techniques

This section discusses various techniques used by dynamic analysis systems for identifying malware and vulnerable applications. Many researchers combine one or more techniques for improving accuracy. Section 6 discusses this in more detail.

- *Virtual machine introspection (VMI)*: VMI reconstructs the emulator's context while working outside of emulator as shown in Fig. 3. VMI achieves this by analyzing relevant kernel data structures, system calls, processes, and extracting valuable information from these data structures. VMI works in contrast to malware detection approaches that run on the same system where malware execute. Root exploit malware after getting administrative privileges can deactivate such a malware analysis system to prevent detection. However, VMI is robust in these cases as monitoring is performed outside of analysis system. Extensive knowledge of

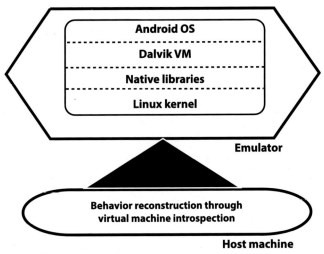

Fig. 3 Virtual machine introspection.

physical hardware and Android OS is required to reconstruct the emulator's context as VMI collects information about all changes happening
in the emulator state because of the execution of given application.

- *API monitoring*: The aim is to get logs of relevant APIs, their run-time
  parameters, return values depending on the objective of the analysis.
  A reflection aware analysis monitors reflection related APIs like
  Class.forName(), Method.invoke(). A leakage detection based analysis
  monitors APIs such as getDeviceId() and sendTextMessage(). The monitoring process can be implemented both at application level [41], [42] or
  at Android OS level [39]. Former approach is scalable and independent
  of API level while the latter approach is more robust against malware that
  check app modifications.

- *Behavior-based monitoring*: These techniques are based on monitoring
  runtime behavior of an app. Runtime information is collected in respect
  of invoked system calls, network access profile, files & memory modifications, URLs accessed, and suspicious API calls. Various tools like
  strace [43], ltrace [44] or their modified versions are used for collecting
  system calls. Tools like PCAP [45], tcpdump [46], and packet sniffers are
  used for collecting network access details. These features are further used
  for classification of apps using machine learning algorithms. MONET
  [47] proposes to use binder transaction information along with system

calls for gathering more accurate semantic information. Researchers are focusing deep learning based classification instead of traditional machine learning to prepare more accurate detection model [48, 49]. Deep learning uses multilayered structure for implementing analysis system, thus perform better classification [50].

- *Taint analysis*: Taint Analysis tracks the information flow out of the device. "Source" and "Sink" are commonly used terms in taint analysis where the Source is any component that takes information of the device such as camera, contacts, emails, location, etc. and the Sink is a component that transfers the information out of the device like network, SMS, emails, etc. Taint Analysis identifies the flow from Source to Sink [51].

- *Forensic analysis*: The forensic analysis in Android is based on analysis of file system, system logs, and physical memory. Such analysis is essential for cyber investigations as it provides information about running processes, terminated processes, open files, network activities, memory mappings, and more. Joe et al. [52] explored issues related to acquiring physical memory captures from Android-based devices and analysis of the acquired data. Authors in [53] presented the case study of real malware from honeynet [54] forensic challenge. Amato et al. [55] proposed methodology for improving results of forensic analysis by using semantic annotations. Semantically annotating evidences make the retrieval of relevant data more effective.

6. Taxonomy of dynamic analysis systems

We comprehensively survey the tools and techniques concentrated on the dynamic analysis of applications in Android. We prepared the taxonomy and grouped analysis techniques into four classes considering objective of the analysis. As shown in Fig. 4, the objectives of dynamic analysis may be to investigate malware (Section 6.1), automating the testing for achieving scalability (Section 6.2), analyzing vulnerabilities in apps (Section 6.3), or similarity detection of apps (Section 6.4). The taxonomy adds one more layer of classification based on the implementation mechanism. A large number of contributions are made in each class by researchers. An exhaustive study of various contributions along with loads of experiments is done. Then, the major contributions in each category are selected for inclusion in this chapter. The selection is done based on relevance, quality, and indexing.

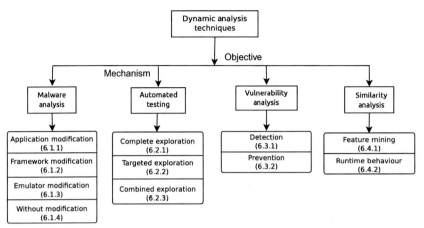

Fig. 4 Dynamic analysis systems taxonomy.

6.1 Dynamic analysis for malware detection

The analysis of apps aims at identifying various malicious activities such as private information leakage, financial loss to the device owner through SMS to premium rate numbers, network access for information leakage or carrying other cyber crimes, device file system alteration, privilege escalation, dynamic code loading (DCL), etc. This section examines security solutions focusing mainly on detection of these malicious activities. As shown in Fig. 4, we classify dynamic malware analysis frameworks into four subcategories based on the approach of implementation as—(1) application modification, (2) framework modification, (3) emulator modification, and (4) without modification. In first, the app under analysis is instrumented/repackaged to make it amenable for analysis. This approach is most deployed approach as it is adaptable being neither API specific nor any OS related modifications are required. The malware may, still check for the presence of changes in the app and may not execute any malicious action. In the second, Android framework is modified to perform monitoring tasks. The main challenge is the in-depth understanding of Android is required for modification. Modification in one version is a not directly mappable to all versions of Android. Also, it must be ensured that modifications do not introduce any vulnerability. The third way is by modifying QEMU [56] based emulators instead of modifying entire Android OS. The benefit of the approach is independence from the Android version. The last approach does not perform modification in any of component. Sections 6.1.1-6.1.4 discuss significant contributions in each subcategory.

6.1.1 Application modification

A. *DroidTrace*: DroidTrace [57] is system-call monitoring based approach that aims to analyze dynamically loaded code with malicious behaviors. It uses `ptrace` [58] to monitor various DCL related system calls of the process under analysis and classifies the application as malicious based on sequence of system calls. It monitors four types of malicious behaviors: file access, network connection, inter-process communication, and privilege escalation. The important contribution of DroidTrace is its forward execution methodology, which automatically triggers most of the dynamically loaded behaviors. Forward execution performs modification in the application to automatically trigger DCL paths. However, malicious activities through other means like SMS, logs are ignored by DroidTrace as it only focuses on DCL paths.

B. *Aurasium*: Aurasium [59] is an application hardening system that intercepts almost every interaction between application and OS through the insertion of monitoring code in the application. The monitoring code alerts the user whenever some sensitive data is being sent out of the device. The interruption is based on policies designed to protect user's privacy and to ensure security. For example, whenever an application attempts to send SMS to premium rate number or accesses some black-listed URL, the monitoring code alerts the user. The technology is widely deployable as it does not need any changes to be made in Android framework. During experiments, we observed that application becomes in-executable after repackaging.

C. *NativeGuard*: NativeGuard [21] focuses on security threats by third-party native libraries. It moves third-party native libraries to an entirely new service application that runs alongside the original (called client) with limited permissions. The two applications interact via interfaces defined by the Android Interface Definition Language (AIDL) and the application holding native libraries acts as a service. It runs in the background and responds to requests from the client. To ensure that the service application starts together with the client and the client is bound to it after being launched. NativeGuard injects crafted dalvik bytecode sequences with the help of `apktool` in app's original bytecode. However, the approach affects runtime behavior as it may restrict some legitimate libraries that do require permissions. The approach is still manual which limits scalability.

D. *Harvester*: Harvester [4] aims malware identification in the presence of obfuscation, reflection, and string encryption. It first performs static backward slicing using call graph and then generate reduced apk using calculated slices. The reduced apk consists of the code necessary to compute Values of Interest, i.e., values of variables at various suspicious points. The reduced apk is then executed in the emulator and run-time logs are collected. Run-time values from logs are injected in the app's bytecode to generate enriched app which can be analyzed by static analyzers. However, slicing is limited to a single Android component, which restricts the detection of inter-component and inter-app leaks. Any missed edge in slicing leads to missed detection of Value of Interest.

6.1.2 Framework modification

A. *StaDynA*: StaDynA [39] complements the static analysis techniques with information obtained from the dynamic analysis. This hybrid solution, first, generates Method Call Graph (MCG) of the app under analysis from its bytecode. It then extends the MCG with the information gathered during dynamic analysis. With additional information obtained during run-time, it can handle DCL and reflection that are missed by existing static analysis frameworks. StaDynA uses manual triggering during dynamic analysis that restricts its applicability for automated analysis. StaDynA supports Android version 4.1.2_r2 that limits the applicability due to version compatibility problems.

B. *BareDroid*: BareDroid [60] uses real devices instead of emulators to make the analysis of antiemulation malware feasible. The biggest problem is restoring the device to initial state after each analysis. BareDroid implements the solution to achieve fast restoring of real devices to their original state. Instead of modifying the kernel or adding specific components, BareDroid leverages SEAndroid [61], which implements the mandatory access control (MAC) model in Android.

BareDroid takes 31.768 s (on average) in comparison to complete device restoration which takes 141.268 s (on average). However, it is less robust due to the usage of hardware devices (phones, USB hubs, and USB cables) which have the higher chances of failure. Also, malfunctioning of a USB hub may lead to quick battery discharge of devices. It is more challenging to scale due to the scheduling of a large number of apps on multiple devices.

C. *TaintDroid*: TaintDroid [62] uses real-time dynamic analysis to capture runtime environment context of the application. It dynamically taints private data and propagates the taint dynamically to detect leakage of private information. TaintDroid implements this by modifying Binder and Dalvik VM. It uses four levels of tracking: variable, message, method, and file to speed up the analysis. TaintDroid cannot monitor the malicious behaviors implemented through native code, covert channels, and control-dependence. TaintDroid's build is available for Android 4.3.1, i.e., API level 18.

D. *XManDroid*: XManDroid [63] extends Android's monitoring mechanism for prevention of privilege escalation attacks at run-time. Privilege escalation attacks exploit vulnerable interfaces of applications (confuse-deputy attacks) or distribute the malicious activity across multiple applications (collusion attacks). XManDroid maintains a system state for the applications installed on the device and communication links (control and data flow) among them. It is invoked whenever Android's default reference monitor grants Inter-Component Communication (ICC) call and validates whether the requested ICC call can potentially be exploited (in combination with other communication links which have ever occurred in the system) based on the system policies. XManDroid can capture leaks implemented through control flows and covert channels in contrast to TaintDroid, which primarily focuses on detecting data flows. However, due to strict policies, XManDroid leads to blocking of even non-sensitive and genuine communication among applications.

E. *NDroid*: NDroid [22] identifies information flow due to native code using virtualization based dynamic taint tracking system. It identifies scenarios of information flows through native code that are missed by other dynamic taint analysis systems. NDdroid is tested by integrating it with TaintDroid. NDroid extends the taint propagation of TaintDroid to native code by storing taint information of native runtime stack. To achieve this, it instruments JNI related functions in Android. However, it does not track control flows. The main limitation of the tool is that, it is not integrated with any advance UI exploration tool which limits scalability and lacks handling antiemulation.

F. *NaClDroid*: NaClDroid [64] is a software fault isolation based framework that embeds Google's Native Client (NaCl) [65] in Android runtime. Native code is compiled using NaCl toolchain and executes in NaCl sandbox. With NaClDroid, malware is not able to read or write memory areas outside of the sandbox. NaClDroid's aim is to protect legitimate applications from malicious libraries. Execution of native code in a separate sandbox prevents various malicious operations performed using native code. Running a native code through NaClDroid causes approximately 10% slowdown on average.

G. *Andrubis*: Andrubis [6] combines static and dynamic analysis at both dalvik VM level and system-level for analyzing malicious applications. Andrubis makes use of Droidbox [66], Taintdroid [62], Androguard [67], and `apktool` [68] for implementation. It modifies the Android framework for VMI, system call monitoring, taint analysis, and API monitoring. It manages antiemulation by using QEMU in single-step mode instead of binary translation (default mode) but due to this, overhead raised by 29%. It cannot analyze files of more than 8 MB size. However, Andrubis relies on the signature of other antivirus tools for classification of the application.

H. *Mobile-Sandbox*: Mobile-Sandbox [69] performs the analysis by monitoring system calls and Java API calls. It monitors both malicious native API calls (outside dalvik VM) and malicious Java calls (within dalvik VM) dynamically. Various features extracted during static and dynamic analysis form an input vector that is given to a linear support vector machine (SVM) [70] to classify the application. The reported limitation is that the modified version of `Ltrace` used for native call tracking kills the application many a times and therefore, the analysis ends. Mobile-Sandbox handles only one or two antiemulation methods. Earlier Mobile-Sandbox was available as online service, but the support has stopped presently.

I. *TraceDroid*: TraceDroid [71] aims to provide better tracer in comparison to the existing Android tracer. TraceDroid modifies Android framework to generate the complete method trace of targeted application. TraceDroid is a hybrid approach where static analysis first collects necessary information by parsing the manifest file and then, the dynamic analysis is performed on emulator. Dynamic analysis uses emulator with modified system image having TraceDroid installed. TraceDroid installs the application and monitors logs, network activities and enables VM tracing. TraceDroid also simulates various system actions such as `BOOT_COMPLETED`, `SMS_RECEIVED`, etc. in addition to Monkey. TraceDroid's experiments reveal that it achieves 50% speedup compared to Android's original profiler. However, TraceDroid does not support tracing of native code. Antiemulation malware can also evade TraceDroid analysis.

J. *AppsPlayground*: AppsPlayground [72] combines detection techniques with automatic exploration techniques. AppsPlayground handles antiemulation partially by implementing techniques that make emulated environment appear as the real device. Malware detection includes multiple approaches like taint tracking, API monitoring, and kernel-level monitoring. Automatic exploration triggers events like BOOT_COMPLETED, SMS_RECEIVED, etc. based on the permissions declared in the application. The fact is that execution of many components depends on these events. AppsPlayground also implements intelligent exploration, which extracts GUI components from application screen and tracks the components based on predefined sequencing policies. To fill the text fields, it does context determination. The nearby strings are used for providing specific strings as input. For example, a textbox is filled with a valid email address if word "Email" is appearing near to textbox. However, AppsPlayground does not include tracing of native code. It is unable to process complex and custom UI widgets.

K. *GlassBox*: GlassBox [73] executes the application in a controlled environment on the real device and extracts various features including Java calls, system calls, and web requests (both encrypted and non-encrypted). These features are used to classify the application as benign or malicious. The approach is paired with the smart monkey, a program developed along with GlassBox that automates the installation, testing of application, and further cleaning of the environment. Currently, the evaluation was done for 100 open source applications selected from F-Droid (Free and Open Source Repository of Android applications) [74]. GlassBox does not support analysis of native code.

6.1.3 Emulator modification

A. *DroidScope*: DroidScope [75] is VMI and taint analysis-based system that reconstructs the semantic behavior of malware at both native and Java level. It collects detailed information of native instructions, dalvik instructions, and API-level activities. It tracks information leaks through both the Java and native components using taint analysis. The entire analysis is performed from outside the VM. DroidScope extended QEMU based Android emulators while the Android system remains unchanged. The limitation is that DroidScope requires manual efforts to analyze the application that makes it inefficient for large-scale analysis.

B. *CopperDroid*: CopperDroid [76] is a VMI based solution for automatic reconstruction of Android malware behavior. CopperDroid's main feature is that it can work for all versions of Android with no modification required for Android system and little change for the emulator. It enhances the QEMU emulator to enable system call tracking. CopperDroid combines system calls and Android specific features observed from binder to detect malware. The features include network access, file system access, SMS, personal information access, and execution of external applications. CopperDroid works for almost all versions of Android including recent version Lollipop. CopperDroid triggers the application artificially with some valid and interesting events such as BOOT_RECEIVED, SMS_RECEIVED based on the app's manifest to increase code coverage. However, CopperDroid is currently available online with its original approach and release of the updated version is pending.

C. *Going native*: Going native [77] proposes a dynamic analysis system to automatically generate security policies and dynamically instruments emulator to record all operations and events executed from native code. The policies allow the normal behavior of native code while restricting the attacks by limiting malicious behaviors like root exploits through native code. An extensive analysis is done for 1.2 million benign applications for differentiating benign and malicious behavior. The results show that the native code sandboxing policy could limit malicious behavior of 99.77% experimented applications.

6.1.4 Without modification

A. *Andlantis*: Andlantis [78] is a scalable framework that runs thousands of Android instances in parallel. It runs on a large commodity cluster of 520 nodes where each node is capable of hosting between 8 and 20 virtual machines. It uses forensic analysis to identify malicious activity. Each application to be analyzed is executed for a specific amount of time in a newly created VM. Then, a filesystem comparison is run against the original Android VM to determine the changes made by the application. The differences are categorized in terms of created, modified, and deleted files. The created and modified files, metadata, and network traffic generated by the app's execution are archived for further analysis. The tool has been tested for Android version 4.2 on x86 architecture only.

B. *CrowDroid*: CrowDroid [79] collects the system call traces from devices of real users based on crowd-sourcing. Through crowd-sourcing, the application is distributed to a large number of users whose interactions with the application will help in the analysis of application. Subsequently, the central server processes the received data and produces feature vectors. It then clusters the feature vectors to create normality model and detects anomalous behavior in Android applications using machine learning. CrowDroid sends only non-personal behavior-related data of each application's user. The battery drainage is very fast due to the continuous fetching of data from devices. However, the data that is sent to the server is not authenticated. This can lead to integrity violation. A single central server may result in scalability challenges.

C. *DroidBox*: Droidbox [66] uses taint analysis and API monitoring. Droidbox records various behavioral information of application under tests such as file operations, telephony operations (SMS and phone calls), cryptographic operations, and traffic monitoring for verifying data leakage. Droidbox generates useful information regarding application execution but lacks automatic result analysis to predict application behavior.

Discussions: Tables 2–5 present the comparison of analysis techniques based on various important features. Column "Version" in all tables shows that only few techniques are independent of Android version. Antiemulation is not relevant for techniques that use on-device (Android device for analysis) analysis. Column "Antiemulation" shows that only Mobile-Sandbox and AppsPlayground handle antiemulation out of all the discussed off-device (emulator for analysis) analysis techniques that too partially. Native code analysis support is not implemented in many techniques as shown in column "Native Code Support." We measured "scalability" based on exploration strategy, average analysis time, and version support. Column "Availability" indicates the availability of source code. Among discussed techniques, TraceDroid, CopperDroid, and Mobile-Sandbox are available as online web-based service. CopperDroid is nonfunctional since long. On-device analysis techniques BareDroid, TaintDroid, CrowDroid require rooted device for analysis, but unfortunately, this makes malware more robust as malware also get root privileges. However, testing based on manual analysis limits the scalability but many approaches are using manual exploration while analyzing application.

Table 2 Comparison of analysis frameworks implemented with application modification.

Tool	Version	Antiemulation	Native code support	Evaluation (#Apps)	Average analysis time	Availability	Scalability	Exploration	Platform
					Property				
DroidTrace	ID	N	Y	50,294	NA	N	Y	Custom	E/D
Aurasium	2.3.6	NR	Y	4751	14–35% overhead	Offline	N	NR	D
NativeGuard	4.3	NR	Y	31	Modest	N	N	Manual	D
Harvester	ID	N	Y*	16,799	3 min	N	Y	Custom	E

N—feature not supported, Y—feature supported, NA—information not available, NR—feature not relevant, ID—version independent, P—feature partially supported, E—emulator, D—device, *—when source and sink are in dex.

Table 3 Comparison of analysis frameworks implemented with framework modification.

Tool	Version	Antiemulation	Native code support	Evaluation (#Apps)	Average analysis time	Availability	Scalability	Exploration	Platform
						Property			
StaDynA	4.1.2	NR	N	10	NA	Offline	N	Manual	D
BareDroid	5.1.0	NR	N	9	NA	Offline	Y	Manual	D
TaintDroid	2.1	NR	Y	30	NA	Offline	N	Manual	D
XManDroid	2.2.1	NR	N	50	NA	Commercial	Y	Monkey	D
NDroid	NA	N	Y	8	NA	Offline	N	Manual	E
NaClDroid	ID	NR	Y	NA	NA	N	N	NA	D
Andrubis	2.3.4	N	Y	1,034,999	10 min	N	Y	Monkey	E
Mobile–Sandbox	4.4	P	Y	10,500	20 min	Online	Y	MonkeyRunner	E
TraceDroid	2.3.4	N	N	492	NA	Online	Y	Monkey Custom	E
AppsPlayground	4.2	P	N	3992	NA	N	Y	Custom	E
Glassbox	4.4 to 6	NR	Y	100	NA	N	Y	Monkey Custom	D

N—feature not supported, Y—feature supported, NA—information not available, NR—feature not relevant, ID—version Independent, P—feature partially supported, E—emulator, D—device.

Table 4 Comparison of analysis frameworks implemented with emulator modification.

Tool	Version	Antiemulation	Native code support	Evaluation (#Apps)	Average analysis time	Availability	Scalability	Exploration	Platform
						Property			
DroidScope	4.3	N	Y	9	NA	Offline	N	Manual	E
CopperDroid	ID	N	Y	2900	10 min	NF	Y	MonkeyRunner + Custom	E
Going Native	4.3	N	Y	1,49,949	NA	N	Y	Monkey	E

N—feature not supported, Y—feature supported, NA—information not available, ID—version independent, E—emulator, D—device, NF—nonfunctional.

Table 5 Comparison of analysis frameworks implemented without modification.

Tool	Version	Antiemulation	Native code support	Evaluation (#Apps)	Average analysis time	Availability	Scalability	Exploration	Platform
						Property			
Andlantis	4.2	N	N	1291	1.16 sec	N	Y	MonkeyRunner	E
CrowDroid	ID	NR	N	20	NA	N	Y	NR	D
DroidBox	ID	N	N	NA	NA	Offline	N	Manual	E

N—feature not supported, NR—feature not required, Y—feature supported, NA—information not available, NR—feature not relevant, ID—version independent, E—emulator, D—device.

6.2 Dynamic analysis for automated testing

Automated and efficient application exploration is very crucial in dynamic analysis because it affects code-coverage, scalability, and overall analysis. Moreover, Android applications have rich functionalities with a large number of different types of views such as scrolls, tabs, drawers, long clicks, etc. The behavior of some applications also depends on system events. A large number of research groups have focused on application exploration specifically. These techniques can be broadly classified into three categories—(1) Complete Exploration that explores the application exhaustively. The method has better coverage but suffers delay in analysis, (2) Targeted Exploration where the selected paths of application are explored based on analysis criteria. The method is fast compared to first and requires complex logic. Targeted exploration techniques make large-scale analysis feasible, and (3) Combined Exploration techniques combine both the methods. This section covers contributions that focus on automated application exploration.

6.2.1 Complete exploration

A. *PUMA*: PUMA [80] targets to enable scalable and programmable UI exploration that can be integrated with various dynamic analysis systems. It is better than random monkey considering systematic exploration. It separates exploration logic from analysis logic. Dynamic analysis tools can use PUMA directly and can focus on analysis. Although it is available offline, it does not provide good documentation of extending and integrating. PUMA is based on Android's UI Automator [81] tool that does not support user-defined widgets. The installation and extension of PUMA is much complex than Monkey.

B. *DynoDroid*: Dynodroid [82] instruments the Android SDK for capturing system events (such as broadcast receiver and service events) and uses Android's Hierarchy Viewer [83] for generating UI events. It observes all the relevant inputs for current state, selects and executes one of the inputs based on predefined strategies. It achieves code coverage of 55% on average. The system is evaluated for two aspects—(1) source code coverage concerning other input generation approaches Monkey and manual execution and (2) discovery of bugs in applications. Dynodroid's results show that Monkey also achieves comparable code coverage, but Monkey requires nearly 20 times more input events for same code coverage. Dynodroid is developed for Android 2.3 version which is very old nowadays. It restricts the application under test to communicate with other applications.

C. *Model-based Android GUI testing*: Baek et al. [84] presented model-based Android GUI testing that works on a GUI model for systematic test generation and efficient debugging support. The authors propose a set of multilevel GUI Comparison Criteria (GUICC) to handle dynamically constructed (custom) GUIs. However, the GUICC selection technique is manual. The evaluation is done for 20 applications so the scalability cannot be assured.

D. *CuriousDroid*: CuriousDroid [85] performs dynamic instrumentation of target application, application layout decomposition, heuristic input generation for intelligent, and user-like exploration. The authors integrated it with the well-known malware analysis sandbox, "Andrubis" and replaced its `Monkey` with CuriousDroid. It has two main components: *UIAnalyzer* and *InputDriver*. The *UIAnalyzer* performs dynamic instrumentation to inject itself into the target application. It is responsible for analyzing the UI, inferring context, and tracking visited activities. The *InputDriver* component executes as a separate process and is responsible for sending user's inputs to the application.

6.2.2 Targeted exploration

A. *Brahmastra*: Brahmastra [86] tests third-party component integration in Android applications. These third-party components may be deeply embedded inside the application and have a higher probability to miss by general GUI testing tools. It statically builds a graph of app's activities and transitions between them. It uses obtained path information to guide the runtime execution toward the third-party components under test. Therefore, run–time performance is guided by path information generated using transition graph. It rewrites the application to trigger navigation to third-party components instead of focusing on GUI Exploration externally. Brahmastra compared itself with PUMA and claims that it achieves 2.7 times improvement over PUMA in invoking target methods.

B. *IntelliDroid*: IntelliDroid [87] takes targeted APIs that can be used for performing malicious activities as input and discovers the event handlers that can trigger the invocation of targeted APIs using call graph-based static analysis method and then generates inputs to these event handlers at runtime. For each targeted API, a target call path is extracted. This path contains the sequence of all the methods beginning from the entry-point to the invocation of the targeted API. IntelliDroid iteratively detects such paths and computes the appropriate inputs to inject, as well as the order in which to inject them. It introduces input at the lower-level device-framework instead on GUI, which makes it possible to integrate with dynamic analysis tools. The authors integrated IntelliDroid with TaintDroid and depicted that the system can detect sensitive data leaks with high scalability. IntelliDroid is unable to compute inputs that are processed by complex functions (e.g., encryption or hashing) in a path constraint. The statically constructed call-graph would be incomplete in the presence of obfuscation.

6.2.3 Combined exploration

A. A^3E: A^3E [88] constructs control flow graph from application bytecode which captures transitions among activities (application screens). This graph is subsequently used in exploration. A^3E implements exploration using—(1) `Targeted` and (2) `Depth-first Exploration (DFS)` exploration. In first, direct exploration of only suspicious activities is performed. Second simulates actions for systematically exploring all activities. Although DFS is slower than targeted exploration, it is more systematic. However, A^3E left multitouch gestures such as pinching and zooming. App's native code cannot be explored as the approach is based on bytecode. Analysis time is more than an hour which limits scalability.

Discussions: Table 6 illustrates the comparison of various automated GUI exploration focused systems. As viewed in column "Version," even few significant contributions like Dynodroid, A^3E, and IntelliDroid are not suitable for recent versions of Android. The column "Integration Support" depicts that only CuriousDroid and IntelliDroid have implemented successful integration with dynamic malware analysis tools. Column "Black-Box" shows that whether tool require source code of for analysis. Dynodroid needs source code of the application for exploration which restricts it for testing of `apk` files available on stores. Android is facilitating very complex and advanced GUI widgets in latest versions. Experiments with various open-source tools conclude that most of the techniques are incompetent in exploring complex and custom widgets. Also, coverage of all possible paths is overwhelming and limits the scalability. Most of the state-of-the-art approaches lack exploration of native code.

6.3 Dynamic analysis for vulnerability analysis

The unintentional faults (vulnerabilities) present in any software makes it the target of exploit(s). Mitra et al. [89] developed benchmark containing samples of 25 known vulnerabilities of Android applications. These vulnerabilities are categorized into six major classes, i.e., ICC, Networking, Cryptographic, Storage, System, and Web. The major vulnerability in

Table 6 Comparison of intelligent GUI exploration focused dynamic systems.

					Property				Integration	
Tool	Version	Code-coverage	Approach	Modification	Evaluation (#Apps)	Platform	Availability	Integration Support	Black-Box
PUMA	4.3	NA	DY	A	3600	E/D	Offline	N	Y
Dynodroid	2.3.5	47%	DY	F	1050	E	N	N	N
GUICC	ID	31% (Activity)	H	W	20	E/D	N	N	Y
CuriousDroid	ID	NA	DY	W	38,872	E/D	N	Y	Y
Brahmastra	ID	34% Hit rate	H	A	1010	E	Offline	N	Y
IntelliDroid	4.3	NA	H	F	NA	E	N	Y	Y
A^3E	2.3.4	59.39–64.11% activity 29.53–36.46% method	H	W	25	D	Offline	N	Y

N—feature not supported, NR—feature not required, Y—feature supported, NA—information not available, W—without, ID—version independent, E—emulator, D—device, F—framework, A—application, H—hybrid, DY—dynamic.

Android is due to unverified ICC. `Intents` without any validation makes the application vulnerable. DCL without validation may lead to remote code injection attacks. Android support for HTML, SQLite, native code, etc., which makes it vulnerable to cross-site scripting (XSS) attacks [90], SQLite based attacks [91], and classical memory corruption attacks [92]. Cryptographic vulnerabilities of applications can be exploited by brute-force attacks, dictionary attacks, chosen-plaintext attacks, etc. The approach for analysis is either detection and reporting of vulnerabilities or prevention of exploitation due to vulnerabilities. Sections 6.3.1–6.3.2 discuss few most relevant contributions.

6.3.1 Detection of vulnerabilities

A. *IntentDroid*: The goal of IntentDroid [93] is to identify Inter-Application Communication (IAC) vulnerabilities. It determined eight vulnerabilities that can arise due to unsafe handling of incoming IAC messages. It monitors all the APIs having source/sink functionalities and run–time data passed through IAC. This information is taken in account for the analysis of vulnerable points. IntentDroid reports possible vulnerable points in the application.

B. *ContentScope*: Content provider is by default accessible to all the components defined within the application. The developer sets read and write permissions for the content provider along with the property "exported" in manifest file of the application. Other applications can access the content provider by requesting the corresponding permission. Android API level 16 and earlier, by default sets "exported" property of the content provider as `true`. Due to this, other applications can access the exported components. ContentScope [94] focuses on detection of vulnerabilities related to content provider that can leak or pollute data. It generates function call graph of an application and checks for the presence of any path from a `public` functions to content provider functions like `ContentProvider.query()`, `db.insert()` using path-sensitive analysis. If such paths are present then, ContentScope automatically derives necessary constraints and prepares appropriate inputs to confirm the vulnerabilities. The prepared inputs are then dynamically fed into the application.

6.3.2 Prevention of vulnerabilities

A. *Grab'n run*: Unverified download of code in DCL can lead to remote code injection attacks. Grab'n Run (GNR) [95] designed a code verification protocol, which implements the secure version of DCL APIs and libraries. For example, it implemented secure *SecureDexClassLoader* API as an extension of *DexClassLoader* API (performs dynamically loading of dex file without verification). *SecureDexClassLoader* wraps *DexClassLoader* with additional security checks. Not only it helps in developing secure DCL implementation in new applications but also provide a repackaging tool, which can patch current insecure real world applications. It is available as Java Open Source Library which is compatible with both ADT and Android Studio. However, currently it only provides secure implementation of *DexClassLoader* API but other APIs like *PathClassLoader* or *android.content.Context. createPackageContext* are also used for dynamic code loading. All dynamically loaded code will be stored in application's private folder by GNR, so it makes inter-app sharing of DCL code challenging even developed by same developers.

6.4 Dynamic analysis for similarity detection

Malware are designed by transforming existing ones using various transformation techniques. The aim is rapid development of malware and also defeat analysis tools based on signature matching techniques. A similarity analysis technique works on finding the variants of known malware in contrast to analyzing individual app. A large number of similarity detection systems based on static analysis [96–100] have been proposed. Although, static detection provides efficient results, however, these contributions will be restricted in the case where repackaged application is protected by obfuscation or hardening. We could find very few research contributions that use dynamic analysis.

6.4.1 Feature mining based similarity detection

A. *ResDroid*: The obfuscation/hardening of app might fail static clone detection techniques based on bytecode comparison as the bytecode might change significantly by hardening. ResDroid [101] is based on the concept that obfuscation tools obfuscate dex files leaving out resource files (layout xml files). The reason of not obfuscating resource files is that resource obfuscation mostly affect the app's functionality or quality-of-experience (QoE). ResDroid first dumps *classes.dex* file from memory of running process since DVM always de-obfuscate files for execution in memory. Then, it extracts statistical and structural features from resource files (part of dumped files) and apply static mining based approaches to quantify applications similarity. ResDroid follows two stage approach: (1) first stage extracts statistical features which are lightweight in terms of extraction and comparison. Based on statistical features, it divides the applications in small groups. (2) second stage extracts structural features within each group for identifying similar applications. Structural features require complex processing but have better classification accuracy. The results of large scale evaluation of ResDroid illustrate that it can identify obfuscated repackaged applications efficiently and effectively.

6.4.2 Runtime behavior based similarity detection

A. *MONET:* The authors in [47] propose MONET, an approach combining runtime behavior with static signatures to detect malware variants. MONET includes backend server and client side application, which can be easily deployed on user's device. MONET has 99% accuracy in detecting malware variants and obfuscated malware. It combines both the system calls and binder transactions for describing runtime behavior. The evaluation depicts that the combination yields more semantic information compared to the systems using only system calls as the matrix for malware classification. MONET's overhead on battery and CPU is minimal.

7. Evaluation

We evaluate the analysis techniques that are available offline or online against the malicious applications. Of 21 papers compared in Tables 2–5, nine were available either as online or offline. However, two frameworks NDroid and BareDroid have some installation issues, and CopperDroid's web-link is non-functional since long time. Therefore, we experiment with remaining six techniques. Online services such as AVC UnDroid [102], NVISO ApkScan [103], AndroTotal [104], VisualThreat [105], VxStream Sandbox [106], OPSWAT Metadefender [107], VirusTotal [108], Cuckoo [109] provide online analysis facility. The approach of these services is not public. VirusTotal is the most widely used service using more than fifty external analyzers for analysis. VisualThreat, NVISO ApkScan, and Cuckoo provide hybrid analysis among these. VisualThreat was earlier provisioning free online analysis but currently, it has stopped the free service. Cuckoo has some issue with dynamic analysis as it always show long queue of samples.

It fails to execute our submitted samples. Therefore, results by NVISO ApkScan are included in this section. Our evaluation aims at answering following research questions:

RQ1 Whether the technique is generic enough to analyze applications independent of API level?

To conduct the evaluation, we develop a malicious application which collects user's IMEI and leaks this through SMS after encryption. This application is compiled with different `minSdkVersion`'s as shown in Table 7. It is found that none of the offline techniques would catch the leak for Android API level higher than 18. Online services based on static analysis are independent of application API level as these do not execute the application. NVISO ApkScan is based on hybrid analysis, however, it cannot analyze applications higher than 16 API level. These results infer that the research community need to focus on the development of the approaches for analyzing applications independent of version. This kind of malware make execution successful on most of the devices using latest API level while evade detection by analysis tools.

RQ2 Whether the technique can detect the leaks employed in dynamically loaded code?

Table 8 depicts the evaluation of tools against DCL. Three different applications were taken from DroidBench [110] for evaluation—(1) Source in DCL, (2) Sink in DCL, and (3) Both Source and Sink in DCL. Only TraceDroid is successful in the identification of leak employed in app's dynamically loaded code. This is the foremost requirement from dynamic analysis as static analysis cannot detect leaks through DCL. Most of the

Table 7 Evaluation against compatibility.

Tool	API 19	API 21	API 23	API 25
	Property			
TraceDroid	Fails to execute application			
DroidScope	Fails to execute application			
TaintDroid	Fails to execute application			
StaDyna	Fails to execute application			
Aurasium	Supports API level $<= 16$			
DroidBox	Supports API level $<= 16$			
NVISO	Supports API level $<= 16$			

Table 8 Evaluation against dynamic code loading.

	Property		
Tool	**Dynamic Source**	**Dynamic Sink**	**Dynamic Both**
TaintDroid	Apps execute but no leak identified		
TraceDroid	Success	Success	Success
StaDyna	Apps execute but no leak identified		
DroidScope	Apps execute but no leak identified		
Aurasium	Execution fails after Repackaging		
DroidBox	Identify DexLoading, File Read, Write but not leak		
NVISO	Apps execute but no leak identified		

Table 9 Evaluation against native-code.

	Property		
Tool	**NativeSource**	**NativeSink**	**NativeIndirect**
DroidScope	Apps execute but no leak identified		
TaintDroid	Apps execute but no leak identified		
TraceDroid	Not support Native Code Analysis		
StaDyna	Not support Native Code Analysis		
Aurasium	Execution fails after Repackaging		
DroidBox	Not support Native Code Analysis		
NVISO	No	Yes	No

online services identify the presence of DCL loading and mark the application suspicious. We have submitted few applications which perform DCL but without any malicious activity. These applications are marked as suspicious by most of the frameworks which show that online services results are based on presence/absence of DCL, not the actual leak. This leads to too many false positives. NVISO does not catch the leaks of DCL.

RQ3 Whether the technique can detect leaks present in native code?

Table 9 presents the results for applications performing leaks through native code. Three different applications with native code were taken from DroidBench [34] for the evaluation. NativeSource application obtains the IMEI in native code and leaks it in Java code. NativeSink application acquires the IMEI in Java code but leaks it in native code. `NativeIndirect` application obtains IMEI in Java, send this to native code and sends back to Java code where the leak is done. Here, the objective of native code is to hinder tainting. Tables 2–5 show that DroidScope, TaintDroid, and Aurasium include handling of native code, however, for experimental samples these missed the detection of native code. VxStream reports that application may leak the data due to presence IMEI read and SMS send permissions present in the manifest file. However, it does not confirm the same from native code analysis. Such permission-based approaches always lead to large false positives which is not a good solution. NVISO ApkScan executed the sample successfully and its response is also fast. However, it is not able to detect all leaks in native code. Overall, based on results, we conclude that leaks hidden in the native code need more attention from research community.

RQ4 Whether the analysis tool is really scalable?

The rapid increase in the number of applications on app markets demands a scalable solution. Scalability of dynamic analysis system depends on factors like automated exploration, device restoration time (in the case of on-device analysis approach). Approaches like CuriousDroid integrates automated GUI exploration with well-known tool Andrubis which will be helpful in providing scalability. Although, CuriousDroid is not formally released as open source tool. During experiments with CuriousDroid, we observe that it is non-functional currently. The same is confirmed by authors also. Online services are scalable due to use of cloud platform but most of these are limited to static analysis.

RQ5 Whether the technique is robust enough to address antiemulation?

TaintDroid, StaDynA, and Aurasium being on-device analysis techniques do not need handling of antiemulation. However, handling of antiemulation is mandatory for techniques using emulated environment for analysis. There are various approaches for checking the presence of emulated environment as detailed in Section 4. Tables 2–5 show that very few off-device techniques handle antiemulation. Research proposals with practical solutions for handling antiemulation are the primary requirement of Android dynamic analysis.

RQ6 Whether the exploration is sufficient to invoke deeply hidden malicious activities?

We constructed a malicious application that performs the malicious action after successful login. The aim is to check that state-of-the-art automated exploration solutions can trigger this malicious behavior or not? TraceDroid uses Android Monkey which lacks intelligent exploration of login and password and thus fails to detect malicious activity. StaDynA and DroidBox require manual analysis, so we provided inputs to the application manually. Both were able to detect the leak but looking at scalability; the approaches need to be extended with automated analysis.

8. Open research issues

After extensive study of field, we are summarizing the features that need special emphasis by the researchers in order to attain an effective solution in Android dynamic analysis.

- *Scalability*: The zillions of applications on various app-stores demand a solution with high scalability. Dependency on manual analysis restrains the scalability of existing approaches. Therefore, antimalware solutions must avoid dependency from any manual analysis. An approach with distributed processing will be favorable in achieving scalability.
- *Finite analysis time*: The majority of current dynamic analysis solutions take a lot of time for analysis. Out of this, a considerable amount of time is required for starting fresh emulator. Approaches for quick restoration of emulator to clean fresh environment will be alluring.
- *Better antivirus*: End users generally use antiviruses on their smartphones but investigations of [7] at large scale show that the majority of antivirus products are severely impacted by even trivial code obfuscations; The study found that antivirus products are purely based on static signature based analysis and are lazy in including new signatures of malicious applications. To deal with obfuscated malware, dynamic analysis with scalable solution in finite time should be incorporated to achieve high accuracy.
- *Advanced antidetection*: Each of the antidetection issues as discussed in Section 4 must be handled efficiently.
- *Real-time analysis*: It is an important research issue to work on productive approaches for on-device data protection. A proactive and real-time analysis approach is desirable that can instantly alert the user on suspicious activity.
- *Improvement in online services*: Online services are mostly based on permission analysis to speed up analysis timings, but this leads to too many false positives. Other important parameters must also be taken care while classifying malware to reduce false positives.
- *Holistic solution*: The majority of the research for Android dynamic analysis focused on isolated areas like several independent solutions for

automated testing have been proposed. However, a holistic solution concerning integration of automated testing with analysis system should be developed.

- *Resiliency with fragmentation*: New versions of Android are released quite often. It is apparent from various tables in Section 6 that very few of solutions work independent of Android version. The dependency on versions will make the analysis system impractical in real.
- *Architecture compatibility*: As covered in Section 2, Android uses both DVM and ART environments for application execution. Analysis approaches must be designed in a way to support both the environments.

9. Concluding remarks

Analysis and detection of malware for Android is a crucial step since Android applications capture user's personal and sensitive information. This chapter enlists various threats to Android users and provides an insight of corresponding solutions. It summarizes techniques for dynamically analyzing different Android malware. Every single day, malware is steadily becoming stealthier by adopting complex antidetection methods. This chapter lays focus on these antidetection methods. The chapter reviews and categorizes various state-of-the-art dynamic analysis techniques along with different challenges posed by each. The chapter furthermore includes an empirical evaluation of the state-of-the-art dynamic analysis approaches using publicly available and self-developed malware. The evaluation shows the limitations of discussed analysis techniques in detection of recent malware as these incorporate several antidetection methods with different levels of complexity. We discuss issues that demand immediate focus of the research community.

We observe that the major concern of dynamic analysis is to achieve scalability. We believe that hybridization of static and dynamic analysis contrary to pure dynamic analysis would definitely improve the performance. A multilayer solution in which a static approach carry out the utmost analysis and further guide dynamic analysis to restrict its analysis to only the portion which can't be managed by static would definitely prove to be scalable solution. Moreover, efficient integration of automated exploration techniques with dynamic analysis frameworks is essential. We believe that techniques which can provide guaranteed exploration of malicious or vulnerable code of application during automated analysis would really help in reducing false negatives. Support for analysis of both Java and native code would play important role in realizing a complete solution. The exponential increase in the number of Android apps and the associated security risks highly demand efficient, scalable, and automated solution.

Acknowledgment

The project is funded in part by CEFIPRA under Grant No. IFC/A/DST-CNRS/2015-01/332.

References

[1] https://www.statista.com/statistics/266136/global-market-share-held-by-smartphone-operating-systems/.

[2] Statista: The Statistics Portal, https://www.statista.com/statistics/281106/number-of-android-app-downloads-from-google-play/.

[3] Q2 2015 Android Malware and Vulnerability Report, http://www.360safe.com/news/2015/07/28/q2-2015-android-malware-and-vulnerability-report/.

[4] S. Rasthofer, S. Arzt, M. Miltenberger, E. Bodden, Harvesting runtime values in Android applications that feature anti-analysis techniques, in: Proceedings of the Annual Symposium on Network and Distributed System Security (NDSS), 2016.

[5] R.A. Apvrille, Obfuscation in android malware and how to fight back, Volume July 2014 of covering the global threat landscape, Virus Bulletin Ltd, 2014, pp. 1–10.

[6] M. Lindorfer, M. Neugschwandtner, L. Weichselbaum, Y. Fratantonio, V. van der Veen, C. Platzer, Andrubis–1,000,000 apps later: a view on current Android malware behaviors, in: Proceedings of the 3rd International Workshop on Building Analysis Datasets and Gathering Experience Returns for Security (BADGERS), 2014.

[7] M. Hammad, J. Garcia, S. Malek, A large-scale empirical study on the effects of code obfuscations on Android apps and anti-malware products, in: ICSE 2018, ACM, 2018.

[8] S.R. Choudhary, A. Gorla, A. Orso, Automated test input generation for Android: are we there yet?(E), in: 2015 30th IEEE/ACM International Conference on Automated Software Engineering (ASE), IEEE, 2015, pp. 429–440.

[9] S. Neuner, V. van der Veen, M. Lindorfer, M. Huber, G. Merzdovnik, E. Weippl, M. Mulazzani, Enter Sandbox: Android Sandbox Comparison, in: Proceedings of the Third Workshop on Mobile Security Technologies (MoST), 2014, 2014.

[10] Android Runtime (ART) and Dalvik, https://source.android.com/devices/tech/dalvik.

[11] Intent, https://developer.android.com/reference/android/content/Intent.html.

[12] Android Binder, https://developer.android.com/reference/android/os/Binder.html.

[13] W. Enck, Defending users against smartphone apps: techniques and future directions, in: International Conference on Information Systems Security, Springer, 2011, pp. 49–70.

[14] D.J.J. Tan, T.-W. Chua, V.L.L. Thing, Securing Android: a survey, taxonomy, and challenges, ACM Comput. Surv. (CSUR) 47 (4) (2015) 58.

[15] P. Faruki, A. Bharmal, V. Laxmi, V. Ganmoor, M.S. Gaur, M. Conti, M. Rajarajan, Android security: a survey of issues, malware penetration, and defenses, IEEE Commun. Surv. Tut. 17 (2) (2015) 998–1022.

[16] UI/Application Exerciser Monkey, http://developer.android.com/tools/help/monkey.html.

[17] K. Tam, A. Feizollah, N.B. Anuar, R. Salleh, L. Cavallaro, The evolution of Android malware and Android analysis techniques, ACM Comput. Surv. (CSUR) 49 (4) (2017) 76.

[18] S. Alam, Z. Qu, R. Riley, Y. Chen, V. Rastogi, DroidNative: automating and optimizing detection of Android native code malware variants, Computers & Security 65 (2017) 230–246.

[19] Virus Profile: Android/Lotoor, https://blog.checkpoint.com/2017/05/17/aprils-wanted-malware/.

[20] Virus Profile: Android/Lotoor, https://home.mcafee.com/virusinfo/virusprofile.aspx?key=9166578.

[21] M. Sun, G. Tan, NativeGuard: protecting Android applications from third-party native libraries, in: Proceedings of the 2014 ACM Conference on Security and Privacy in Wireless & Mobile Networks, ACM, 2014, pp. 165–176.

[22] C. Qian, X. Luo, Y. Shao, A.T. Chan, On tracking information flows through JNI in Android applications, in: 2014 44th Annual IEEE/IFIP International Conference on Dependable Systems and Networks, IEEE, 2014, pp. 180–191.

[23] M. Spreitzenbarth, F. Freiling, F. Echtler, T. Schreck, J. Hoffmann, Mobile-sandbox: having a deeper look into Android applications, in: Proceedings of the 28th Annual ACM Symposium on Applied Computing, ACM, 2013, pp. 1808–1815.

[24] T. Petsas, G. Voyatzis, E. Athanasopoulos, M. Polychronakis, S. Ioannidis, Rage against the virtual machine: hindering dynamic analysis of Android malware, in: Proceedings of the Seventh European Workshop on System Security, ACM, 2014, p. 5.

[25] J. Gajrani, J. Sarswat, M. Tripathi, V. Laxmi, M.S. Gaur, M. Conti, A robust dynamic analysis system preventing SandBox detection by Android malware, in: Proceedings of the 8th International Conference on Security of Information and Networks, ACM, 2015, pp. 290–295.

[26] Y. Jing, Z. Zhao, G.-J. Ahn, H. Hu, Morpheus: automatically generating heuristics to detect Android emulators, in: Proceedings of the 30th Annual Computer Security Applications Conference, ACM, 2014, pp. 216–225.

[27] W. Hu, Z. Xiao, Guess Where I Am-Android: Detection and Prevention of Emulator Evading on Android, HitCon, 2014.

[28] R. Vollmer, "Xposed framework", http://repo.xposed.info/module/de.robv.android.xposed.installer.

[29] Cydia Substrate, http://www.cydiasubstrate.com/.

[30] adbi - The Android Dynamic Binary Instrumentation Toolkit, https://github.com/crmulliner/adbi.

[31] A. Bulazel, B. Yener, A survey on automated dynamic malware analysis evasion and counter-evasion: PC, mobile, and web, in: Proceedings of the 1st Reversing and Offensive-oriented Trends Symposium, ACM, 2017, p. 2.

[32] MonkeyRunner, http://developer.android.com/tools/help/MonkeyRunner.html.

[33] A. Kovacheva, Efficient code obfuscation for Android, in: International Conference on Advances in Information Technology, Springer, 2013, pp. 104–119.

[34] DroidBench–Self Modification, https://github.com/secure-software-engineering/DroidBench/tree/develop#self-modification.

[35] Y. Fratantonio, A. Bianchi, W. Robertson, E. Kirda, C. Kruegel, G. Vigna, Triggerscope: towards detecting logic bombs in Android applications, in: 2016 IEEE Symposium on Security and Privacy (SP), IEEE, 2016, pp. 377–396.

[36] C. Marforio, H. Ritzdorf, A. Francillon, S. Capkun, Analysis of the communication between colluding applications on modern smartphones, in: Proceedings of the 28th Annual Computer Security Applications Conference, ACM, 2012, pp. 51–60.

[37] J. Gajrani, L. Li, V. Laxmi, M. Tripathi, M.S. Gaur, M. Conti, POSTER: detection of information leaks via reflection in Android apps, in: Proceedings of the 2017 ACM on Asia Conference on Computer and Communications Security, ACM, 2017, pp. 911–913.

[38] J. Gajrani, V. Laxmi, M. Tripathi, M.S. Gaur, D.R. Sharma, A. Zemmari, M. Mosbah, M. Conti, Unraveling reflection induced sensitive leaks in Android apps, in: International Conference on Risks and Security of Internet and Systems, Springer, 2017, pp. 49–65.

[39] Y. Zhauniarovich, M. Ahmad, O. Gadyatskaya, B. Crispo, F. Massacci, StaDynA: addressing the problem of dynamic code updates in the security analysis of Android applications. in: Proceedings of the 5th ACM Conference on Data and Application Security and Privacy, ACM, New York, NY, USA. ISBN: 978-1-4503-3191-3, 2015, pp. 37–48, https://doi.org/10.1145/2699026.2699105.

[40] H. Han, R. Li, J. Tang, Identify and inspect libraries in Android applications, Wirel. Pers. Commun. 103 (1) (2018) 491–503.

[41] APIMonitor, https://code.google.com/p/droidbox/wiki/APIMonitor.

[42] D.-J. Wu, C.-H. Mao, T.-E. Wei, H.-M. Lee, K.-P. Wu, Droidmat: Android malware detection through manifest and API calls tracing, in: 2012 Seventh Asia Joint Conference on Information Security (Asia JCIS), IEEE, 2012, pp. 62–69.

[43] Strace, http://man7.org/linux/man-pages/man1/strace.1.html.

[44] Ltrace, http://man7.org/linux/man-pages/man1/ltrace.1.html.

[45] Programming With Pcap, www.tcpdump.org/pcap.html.

[46] Welcome to Android tcpdump, http://www.androidtcpdump.com/.

[47] M. Sun, X. Li, J.C. Lui, R.T. Ma, Z. Liang, Monet: a user-oriented behavior-based malware variants detection system for Android, IEEE Trans. Inf. Foren. Secur. 12 (5) (2017) 1103–1112.

[48] D. Li, Z. Wang, Y. Xue, Fine-grained Android malware detection based on deep learning, in: 2018 IEEE Conference on Communications and Network Security (CNS), IEEE, 2018, pp. 1–2.

[49] K. Xu, Y. Li, R.H. Deng, K. Chen, DeepRefiner: multi-layer Android malware detection system applying deep neural networks, in: 2018 IEEE European Symposium on Security and Privacy (EuroS&P), IEEE, 2018.

[50] E.B. Karbab, M. Debbabi, A. Derhab, D. Mouheb, MalDozer: automatic framework for Android malware detection using deep learning, Digit. Invest. 24 (2018) S48–S59.

[51] M. Egele, T. Scholte, E. Kirda, C. Kruegel, A survey on automated dynamic malware-analysis techniques and tools, ACM Comput. Surv. (CSUR) 44 (2) (2012) 6.

[52] J. Sylve, A. Case, L. Marziale, G.G. Richard, Acquisition and analysis of volatile memory from Android devices, Digit. Invest. 8 (3) (2012) 175–184.

[53] J. Li, D. Gu, Y. Luo, Android malware forensics: reconstruction of malicious events, in: 2012 32nd International Conference on Distributed Computing Systems Workshops (ICDCSW), IEEE, 2012, pp. 552–558.

[54] Forensic Challenge 9–"Mobile Malware", http://www.honeynet.org/node/751.

[55] F. Amato, L. Barolli, G. Cozzolino, A. Mazzeo, F. Moscato, Improving results of forensics analysis by semantic-based suggestion system, in: International Conference on Emerging Internetworking, Data & Web Technologies, Springer, 2018, pp. 956–967.

[56] QEMU, https://www.qemu.org/.

[57] M. Zheng, M. Sun, J.C.S. Lui, DroidTrace: a ptrace based Android dynamic analysis system with forward execution capability, in: International Wireless Communications and Mobile Computing Conference, IWCMC 2014, Nicosia, Cyprus, August 4–8, 2014, 2014, pp. 128–133, https://doi.org/10.1109/IWCMC.2014.6906344.

[58] Ptrace: Linux Programmer's Manual, http://man7.org/linux/man-pages/man2/ptrace.2.html.

[59] R. Xu, H. Saïdi, R. Anderson, Aurasium: practical policy enforcement for Android applications, in: Presented as part of the 21st USENIX Security Symposium (USENIX Security 12), 2012, pp. 539–552.

[60] S. Mutti, Y. Fratantonio, A. Bianchi, L. Invernizzi, J. Corbetta, D. Kirat, C. Kruegel, G. Vigna, BareDroid: large-scale analysis of Android apps on real devices, in: Proceedings of the 31st Annual Computer Security Applications Conference, ACM, New York, NY, USA. ISBN: 978-1-4503-3682-6, 2015, pp. 71–80, https://doi.org/10.1145/2818000.2818036.

[61] S. Smalley, R. Craig, Security enhanced (SE) Android: bringing flexible MAC to Android, in: NDSSThe Internet Society, 2013, http://dblp.uni-trier.de/db/conf/ndss/ndss2013.html#SmalleyC13.

[62] W. Enck, P. Gilbert, S. Han, V. Tendulkar, B.-G. Chun, L.P. Cox, J. Jung, P. McDaniel, A.N. Sheth, TaintDroid: an information-flow tracking system for real-time privacy monitoring on smartphones, ACM Trans. Comput. Syst. 32 (2) (2014) 5:1–5:29, https://doi.org/10.1145/2619091.

[63] S. Bugiel, L. Davi, A. Dmitrienko, T. Fischer, A.-R. Sadeghi, Xmandroid: a new Android evolution to mitigate privilege escalation attacks, Technische Universität Darmstadt, Technical Report TR-2011-04, 2011, tech. rep.

[64] G.P. Elias Athanasopoulos, Vasileios P. Kemerlis, A.D. Keromytis, NaClDroid: native code isolation for Android applications, in: European Symposium on Research in Computer Security (ESORICS), Springer, 2016.

[65] B. Yee, D. Sehr, G. Dardyk, J.B. Chen, R. Muth, T. Ormandy, S. Okasaka, N. Narula, N. Fullagar, Native client: a sandbox for portable, untrusted x86 native code, in: 2009 30th IEEE Symposium on Security and Privacy, IEEE, 2009, pp. 79–93.

[66] Dynamic analysis of Android apps, https://github.com/pjlantz/droidbox.

[67] Androguard, https://github.com/androguard/androguard.

[68] Apktool, https://ibotpeaches.github.io/Apktool/.

[69] M. Spreitzenbarth, T. Schreck, F. Echtler, D. Arp, J. Hoffmann, Mobile-Sandbox: combining static and dynamic analysis with machine-learning techniques. Int. J. Inf. Secur. 14 (2) (2014) 141–153. ISSN: 1615-5270, https://doi.org/10.1007/s10207-014-0250-0.

[70] Introduction to Support Vector Machines, http://docs.opencv.org/2.4/doc/tutorials/ml/introduction_to_svm/introduction_to_svm.html.

[71] V. van der Veen, Dynamic Analysis of Android Malware, Ph.D. thesis, VU University Amsterdam 2013, http://tracedroid.few.vu.nl/.

[72] V. Rastogi, Y. Chen, W. Enck, AppsPlayground: automatic security analysis of smartphone applications, in: Proceedings of the third ACM Conference on Data and Application Security and Privacy, ACM, 2013, pp. 209–220.

[73] P. Irolla, E. Filiol, Glassbox: dynamic analysis platform for malware android applications on real devices, in: CoRR, 2016, abs/1609.04718. http://arxiv.org/abs/1609.04718.

[74] F-Droid, https://f-droid.org/.

[75] L.K. Yan, H. Yin, DroidScope: seamlessly reconstructing the OS and Dalvik semantic views for dynamic Android malware analysis, in: Proceedings of the 21st USENIX Conference on Security Symposium, USENIX Association, Berkeley, CA, USA, 2012, p. 29. http://dl.acm.org/citation.cfm?id=2362793.236282229.

[76] K. Tam, S.J. Khan, A. Fattori, L. Cavallaro, CopperDroid: automatic reconstruction of Android malware bBehaviors, in: NDSSThe Internet Society, 2015. http://dblp.uni-trier.de/db/conf/ndss/ndss2015.html#TamKFC15.

[77] V. Afonso, A. Bianchi, Y. Fratantonio, A. Doupé, M. Polino, P. de Geus, C. Kruegel, G. Vigna, Going native: using a large-scale analysis of Android apps to create a practical native-code sandboxing policy, in: Proceedings of the Annual Symposium on Network and Distributed System Security (NDSS), 2016.

[78] M. Bierma, E. Gustafson, J. Erickson, D. Fritz, Y.R. Choe, Andlantis: large-scale Android dynamic analysis, No. SAND2014-1596C, Sandia National Lab. (SNL-CA), Livermore, CA, United States, 2014.

[79] I. Burguera, U. Zurutuza, S. Nadjm-Tehrani, Crowdroid: behavior-based malware detection system for Android, in: Proceedings of the 1st ACM Workshop on Security and Privacy in Smartphones and Mobile Devices, ACM, 2011, pp. 15–26.

[80] S. Hao, B. Liu, S. Nath, W.G.J. Halfond, R. Govindan, PUMA: programmable UI-automation for large-scale dynamic analysis of mobile apps. in: Proceedings of the 12th Annual International Conference on Mobile Systems, Applications, and

Services, ACM, New York, NY, USA. ISBN: 978-1-4503-2793-0, 2014, pp. 204–217, https://doi.org/10.1145/2594368.2594390.

[81] UI Automator, https://developer.android.com/training/testing/ui-automator.html.

[82] A. Machiry, R. Tahiliani, M. Naik, Dynodroid: an input generation system for Android apps. in: Proceedings of the 2013 9th Joint Meeting on Foundations of Software Engineering, ACM, New York, NY, USA. ISBN: 978-1-4503-2237-9, 2013, pp. 224–234, https://doi.org/10.1145/2491411.2491450.

[83] Viewer, http://www.androiddocs.com/tools/help/hierarchy-viewer.html.

[84] Y.-M. Baek, D.-H. Bae, Automated model-based Android GUI testing using multi-level GUI comparison criteria, in: 2016 31st IEEE/ACM International Conference on Automated Software Engineering (ASE), IEEE, 2016, pp. 238–249.

[85] P. Carter, C. Mulliner, M. Lindorfer, W. Robertson, E. Kirda, CuriousDroid: automated user interface interaction for Android application analysis sandboxes, in: Financial Cryptography and Data Security-20th International Conference, FC, 2016.

[86] R. Bhoraskar, S. Han, J. Jeon, T. Azim, S. Chen, J. Jung, S. Nath, R. Wang, D. Wetherall, Brahmastra: driving apps to test the security of third-party components, in: Proceedings of the 23rd USENIX Conference on Security Symposium, USENIX Association, Berkeley, CA, USA. ISBN: 978-1-931971-15-7, 2014, pp. 1021–1036. http://dl.acm.org/citation.cfm?id=2671225.2671290.

[87] M.Y. Wong, D. Lie, IntelliDroid: a targeted input generator for the dynamic analysis of Android malware, in: NDSS, vol. 16, 2016, pp. 21–24.

[88] T. Azim, I. Neamtiu, Targeted and depth-first exploration for systematic testing of Android apps, in: Acm Sigplan Notices, vol. 48, ACM, 2013, pp. 641–660.

[89] J. Mitra, V.-P. Ranganath, Ghera: a repository of Android app vulnerability benchmarks, in: Proceedings of the 13th International Conference on Predictive Models and Data Analytics in Software Engineering, ACM, 2017, pp. 43–52.

[90] W. Bao, W. Yao, M. Zong, D. Wang, Cross-site scripting attacks on Android hybrid applications, in: Proceedings of the 2017 International Conference on Cryptography, Security and Privacy, ACM, 2017, pp. 56–61.

[91] Android SQLite based attacks, http://blog.watchfire.com/files/androidsqlite journal.pdf.

[92] F.H. Shezan, S.F. Afroze, A. Iqbal, Vulnerability detection in recent Android apps: an empirical study, in: 2017 International Conference on Networking, Systems and Security (NSysS), IEEE, 2017, pp. 55–63.

[93] R. Hay, O. Tripp, M. Pistoia, Dynamic detection of inter-application communication vulnerabilities in Android. in: Proceedings of the 2015 International Symposium on Software Testing and Analysis, ACM, New York, NY, USA. ISBN: 978-1-4503-3620-8, 2015, pp. 118–128, https://doi.org/10.1145/2771783.2771800.

[94] Y.Z.X. Jiang, Detecting passive content leaks and pollution in Android applications, in: Proceedings of the 20th Network and Distributed System Security Symposium (NDSS), 2013.

[95] L. Falsina, Y. Fratantonio, S. Zanero, C. Kruegel, G. Vigna, F. Maggi, Grab 'N run: secure and practical dynamic code loading for Android applications. in: Proceedings of the 31st Annual Computer Security Applications Conference, ACM, New York, NY, USA. ISBN: 978-1-4503-3682-6, 2015, pp. 201–210, https://doi.org/10.1145/2818000.2818042.

[96] S. Hanna, L. Huang, E. Wu, S. Li, C. Chen, D. Song, Juxtapp: a scalable system for detecting code reuse among Android applications, in: International Conference on Detection of Intrusions and Malware, and Vulnerability Assessment, Springer, 2012, pp. 62–81.

[97] F. Zhang, H. Huang, S. Zhu, D. Wu, P. Liu, ViewDroid: towards obfuscation-resilient mobile application repackaging detection, in: Proceedings of the 2014 ACM Conference on Security and Privacy in Wireless & Mobile Networks, ACM, 2014, pp. 25–36.

[98] J. Crussell, C. Gibler, H. Chen, Attack of the clones: detecting cloned applications on Android markets, in: European Symposium on Research in Computer Security, Springer, 2012, pp. 37–54.

[99] W. Zhou, Y. Zhou, X. Jiang, P. Ning, Detecting repackaged smartphone applications in third-party Android marketplaces, in: Proceedings of the second ACM conference on Data and Application Security and Privacy, ACM, 2012, pp. 317–326.

[100] M. Fan, J. Liu, X. Luo, K. Chen, Z. Tian, Q. Zheng, T. Liu, Android malware familial classification and representative sample selection via frequent subgraph analysis, IEEE Trans. Inf. Foren. Secur. 13 (2018) 1890–1905.

[101] Y. Shao, X. Luo, C. Qian, P. Zhu, L. Zhang, Towards a scalable resource-driven approach for detecting repackaged Android applications. in: Proceedings of the 30th Annual Computer Security Applications Conference, ACM, New York, NY, USA. ISBN: 978-1-4503-3005-3, 2014, pp. 56–65, https://doi.org/10.1145/2664243.2664275.

[102] AVC UnDroid, http://undroid.av-comparatives.info/about.php.

[103] NVISO ApkScan, https://apkscan.nviso.be/.

[104] AndroTotal, https://andrototal.org/.

[105] Visual Threat, http://www.visualthreat.com/index.action.

[106] VxStream Sandbox, https://www.reverse.it/.

[107] OPSWAT Metadefender, https://www.metadefender.com/.

[108] VirusTotal, https://www.virustotal.com/.

[109] Cuckoo, https://sandbox.pikker.ee/.

[110] DroidBench Dynamic Code Loading, https://github.com/secure-software-engineering/DroidBench/blob/develop/apk/DynamicLoading/.

About the authors

Jyoti Gajrani received the M.Tech. degree in Computer Engineering from the Indian Institute of Technology, Bombay in 2013. She is currently pursuing the Ph.D. degree (Part-time) in computer science with Malaviya National Institute of Technology, Jaipur, under the supervision of Prof. Manoj Singh Gaur from Indian Institute of Technolgy, Jammu and Dr. Meenakshi Tripathi from Malaviya National Institute of Technology, Jaipur. She is working as Assistant Professor (Full-time) in Computer Engineering department at Govt. Engineering College, Ajmer since 2006. Her research interests include the area of security and privacy, with a special emphasis on security in Android.

Prof. Vijay Laxmi received her Master's in Computer Science and Engg. from Indian Institute of Technology Delhi and Ph.D. from University of Southampton, United Kingdom. She has been a faculty in Department of Computer Science and Engineering, Malaviya National Institute of Technology Jaipur, India. Her research interests include Information security, Malware analysis, Security and QoS provisioning in wireless Networks.

Meenakshi Tripathi is an Associate Professor at Malaviya National Institute of Technology (MNIT), Jaipur (India). She obtained her Ph.D. from MNIT in 2015. Her research areas are Wireless Sensor Networks (WSN), Software Defined Networks (SDN) & Internet of Things (IoT). She has published more than 30 research papers in refereed International & National Journals & proceedings of National & International conferences. She has presented her research ideas by participating in various conferences within India & outside India like China, Canada, and United States etc. She is working on various research projects funded by the Department of Science and Technology, Ministry of Electronics and Information Technology, India of worth around Rs. 50 Lakhs. She is the reviewer for several renowned journals such as International Journal of Communication System (Wiley), Security and Communication Networks (Hindawi), Wireless Communications, etc. She has also been part of several International conferences such as SPACE 2015, ICISS 2016, ICSP 2019 etc. She is an upright member of institute of Electrical & Electronics Engineers (IEEE) & Association of Computing Machinery (ACM). She is also the Chairman of Computer Society of India (Jaipur Chapter).

Prof. Manoj Singh Gaur completed his Master's degree in Computer Science and Engineering from Indian Institute of Science Bangalore, India and Ph.D. from University of Southampton, United Kingdom. Prof. Gaur has been a faculty in Department of Computer Science and Engineering, Malaviya National Institute of Technology Jaipur, India and currently Director, IIT Jammu. His research areas include Networks-on-Chip, Computer and network security, Multimedia streaming in wireless networks.

Dr. Akka Zemmari has received his Ph.D. degree from the University of Bordeaux, France, in 2000. He is an Associate Professor in computer science since 2001 at University of Bordeaux, France. He has more than 100 research articles published in peer reviewed Journals and International Conferences, and also serves as programme committee member in the International Conferences. His research interests include distributed algorithms and systems, graphs, randomized algorithms, machine learning and security.

Prof. Dr. Mohamed Mosbah is a Professor in computer science at the Polytechnic Institute of Bordeaux (a high graduate engineering school) and a holder of a bonus for scientific excellence. He is currently the Director of Industrial Partnerships and Innovation in Computer Science and Engineering. He carries his research in LaBRI, a research Lab. in computer science common with the University of Bordeaux and CNRS, where he is currently the Deputy Director. His research areas include distributed systems and algorithms, simulation tools, security protocols, VANET and wireless networks. In particular, he is leading a project over the last years to develop a new model together with an integrated methodological framework for distributed algorithms. In addition to capturing classical distributed systems concepts, this framework provides methods and software tools to design, prove and implement distributed algorithms and protocols. This platform is used to teach courses in distributed computing for Graduate computer science students, and to test and prototype algorithms. He wrote more than 60 articles and developed software tools, and he is involved in various technical program committees and organizations of many international conferences. He is also an expert reviewer of national and European projects. He has directed over 52 Master's theses and over 17 Ph.D. dissertations. Prof. Mosbah obtained a French-Italian Grant in the PHC Galilee Program (2014) on "Secure Distributed Protocols in Wireless Environments". He also obtained two grants from the French Ministry of Industry for collaborative projects with Aeronautics Companies: RECORDS (2008–2010) and SIMID (2010–2013) dealing with the safe distributed information processing. He is involved in the European–Indian REACH project (2015–2018), and he is currently leading a French Tunisian project DERCA-VIe(2017–2019) on "Deployment of wireless sensor networks in the context of Smart Cities".

Mauro Conti is Full Professor at the University of Padua, Italy, and Affiliate Professor at the University of Washington, Seattle, United States. He obtained his Ph. D. from Sapienza University of Rome, Italy, in 2009. After his Ph.D., he was a Postdoc Researcher at Vrije Universiteit Amsterdam, The Netherlands. In 2011 he joined as Assistant Professor the University of Padua, where he became Associate Professor in 2015 and Full Professor in 2018. He has been Visiting Researcher at GMU (2008, 2016), UCLA (2010), UCI (2012, 2013, 2014, 2017), TU Darmstadt (2013), UF (2015), and FIU (2015, 2016, 2018). He has been awarded with a Marie Curie Fellowship (2012) by the European Commission and with a Fellowship by the German DAAD (2013). His research is also funded by companies, including Cisco, Intel, and Huawei. His main research interest is in the area of security and privacy. In this area, he published more than 250 papers in topmost international peer-reviewed journals and conference. He is Area Editor-in-Chief for IEEE Communications Surveys & Tutorials, and Associate Editor for several journals, including IEEE Communications Surveys & Tutorials, IEEE Transactions on Information Forensics and Security, IEEE Transactions on Dependable and Secure Computing, and IEEE Transactions on Network and Service Management. He was Program Chair for TRUST 2015, ICISS 2016, WiSec 2017, and General Chair for SecureComm 2012 and ACM SACMAT 2013. He is Senior Member of the IEEE.

CHAPTER FOUR

Eyeing the patterns: Data visualization using doubly-seriated color heatmaps

Matthew Lane[a,b], Alberto Maiocco[b], Sanjiv K. Bhatia[b], Sharlee Climer[b]
[a]Bayer Crop Science, St. Louis, MO, United States
[b]University of Missouri—St. Louis, St. Louis, MO, United States

Contents

Abstract

In current times, data are being generated at an ever-increasing rate and researchers are challenged to extract meaningful information from this large volume of data. The information extraction can be facilitated by visualization of data, which can be invaluable for deepening insights about trends and exceptions and may facilitate the discovery of patterns, outliers, or other anomalies or characteristics. These visual depictions are provided using various tools, including heatmaps. Heatmaps are matrices of colored cells that represent underlying numeric data. In a typical heatmap, each row represents an object, each column represents a condition, time point, instance, or other property, and the color of each cell indicates the associated data value. However, observations of randomly ordered data are rarely enlightening and rearrangement of rows and columns into clusters of similar data has proven to be of great value. A number of approaches have been developed to tackle this combinatorial problem, including Bond Energy Algorithm, the Traveling Salesman Problem Model, TSP + k, and Hierarchical Clustering. Identifying optimal solutions for the first three of these approaches is NP-hard and the fourth requires exponential computation time. Despite these computational demands, optimality is sometimes pursued. However, approximate algorithms have been developed to address the need for more efficient tools. These approximation techniques are widely varied in both computation time and quality of results. Importantly, even for optimal solvers, these approaches carry assumptions and biases, some of which are quite subtle and commonly overlooked, yet may impact results in significant ways. In short, the choice of rearrangement method should be mindful of the particular characteristics of the data at hand. Another issue in heatmap construction is the failure to properly preprocess data prior to rearrangement. In this chapter, we summarize the history of heatmaps, scrutinize various aspects of data preprocessing, and then examine several algorithms to reorder data rows and columns such that similar data are clustered together. More specifically, we review Bond Energy Algorithm, the Traveling Salesman Problem Model, TSP + k, and Hierarchical Clustering, noting assumptions, strengths, and weaknesses of each approach. This chapter concludes with thoughts for potential future directions of heatmap research.

1. Introduction

The current times are characterized by a proliferation of data in a number of fields. We see continuous production, with the units of large amounts of data moving from exabytes to zettabytes. So much so that the International Bureau of Weights and Measures has recently introduced new metrics of ronnabytes (10^{27}) and queccabytes (10^{30}) to account for the global growth in data storage requirements [1].

The explosive data growth is not just confined to scientific ventures such as Large Hedron Collider but is pervasive across multiple applications including business (e-commerce, transactions, and stocks), science (remote sensing, bioinformatics, and scientific simulations), and society (news, blogs, and social networks). As an example, consider a warehouse for an online store that needs to improve the efficiency of the packing process. This may be achieved by physically placing the items closer to each other on the shelves if those items are commonly ordered together. The information will have to be extracted from orders received. The inventory manager may have access to records for hundreds of thousands of orders but drawing useful information from those orders can be an ordeal. For example, someone will have to recognize that out of hundreds of thousands of orders, a few thousands contain some of those items individually, and a few hundred contain those items together. As a concrete example, an enthusiast will order some expensive lenses for a DSLR camera and invariably, also order some ultraviolet filters to enhance the quality of images taken using those lenses as well as to protect the investment in lenses. Further, the order may also include a lens hood to guard against stray light. It is obvious that the lens can be ordered by itself and the filter and hood may be ordered later on (about 40% of the time), or the items may be ordered together (about 60% of the time). Now, the problem is not only to find the items in the orders individually but also to find the items that have a correlation in being ordered together. In a different scenario, suppose that genetics researchers strive to understand the biological processes underlying Alzheimer's disease. They may have access to data showing expression levels of tens of thousands of genes for thousands of people suffering from the disease. However, the challenge is in finding the pattern in the gene expressions that is associated with the progression of the disease. Again, finding patterns in such a huge matrix of numbers is an extremely hard problem.

The voluminous data has led to new algorithms to recognize patterns in the large amount of collected data. These algorithms have been used in machine learning, and mathematical and statistical data analysis. The large dimensions of data are generally tackled with techniques such as cluster analysis. Cluster analysis is used to discover patterns in data and utilizes techniques such as heatmaps to visually present the data [2]. A heatmap or a shading matrix provides a different color representation for different values in the matrix that is easy to visualize graphically.

This chapter describes the use of heatmaps to visualize patterns in a large amount of data. In the next section, we explore the history of heatmaps from ancient times to the present day. We then dig into several data preprocessing topics, including database input errors and database interaction errors, along with tips for cleaning data effectively. We follow with an examination of several data rearrangement algorithms including Bond Energy Algorithm, the Traveling Salesman Model, TSP + k, and Hierarchical Clustering. We conclude the chapter with speculations of future directions within the heatmap research domain.

2. Background and history

Humans have been collecting data and formatting it for visualization for millennia. Data formatted into columns and rows can be traced back to ancient Sumer c. 2028 BCE, where data rows represent a particular type of animal, and column headers, as well as row titles, along with personal information are detailed [3].

The earliest known published instance of a heatmap is from 1873, found in Loua's *Atlas Statistique de la Population de Paris*, a statistical work in which population data of municipalities was logged and shaded by hand [2,4].

The shading and coloring of individual cells, rows, and columns offers a lot of insight into data. However, we can gain greater amounts of insight through the reordering of matrix data. The reordering of rows with shaded data can offer some insight, and in 1967 Jacques Bertin illustrated a *reorderable matrix* (done so by hand), such that not only the rows were reordered, but the columns as well [2]. For example, assume you have a given binary matrix. By reordering the columns and rows of the matrix, it is possible to visualize patterns more readily than what could be seen originally (Fig. 1).

By rearranging the columns and rows, we can gain a significantly greater understanding of the data by observing natural communities of interrelated items, a phenomenon known as *rearrangement clustering*. Rearrangement clustering has existed under many names such as *matrix reordering*, *structuring of matrices*, and *restricted partitioning* [5].

The rearrangement of data in a set such that comparable datapoints of the same dimension are nearer to each other in the matrix display is known as *seriation* [6]. While it's not always the case that all seriation is done with any particular "distance measure" in mind (e.g. Bertin's rearrangement by hand), most seriation does account for some distance measure in its rearrangement of the matrix. Choices of these distance measures do affect the ultimate result of how columns and rows seriate, and can impact the clustering results [5].

It was not until 1899 that an algorithm for permuting both row and column matrix data was developed by Sir William Petrie [2]. Petrie rearranged artifacts from dig sites in a matrix such that the largest artifacts were along the diagonal of the matrix [7]. Petrie's algorithm for realigning data along the matrix is what would eventually become known as an example of seriation [2]. In his findings, some of Petrie's matrices were inadvertently Toeplitz

Fig. 1 An example of Bertin's reorderable matrix where both rows and columns are reordered.

$$a \quad b \quad c$$
$$d \quad a \quad b$$
$$e \quad d \quad a$$

Fig. 2 An example of a Toeplitz Matrix [7a].

Fig. 3 An example of a Guttman scalogram.

matrices, which are $n \times n$ matrices where the diagonal has constants from left to right as shown in Fig. 2.

Around the same time period, Jan Czekanowski, a Polish anthropologist, created a seriation algorithm by replacing numbers with characters of *differentiated diameter*, ultimately creating a shaded gradient for displaying the matrix data. Czekanowski's similarity coefficient for determining how to seriate the data was the average of physical difference values in his test subjects [8].

In 1950, Robert Guttman created a *scalogram* which was a binary matrix permuted by hand. As opposed to previous iterations of matrix seriation, instead of creating a matrix with a diagonal line around which data would fit, the Guttman's scalogram ordered data in such a way to create a demarcation in the matrix [2,9]. A Guttman's scalogram is presented in Fig. 3.

On top of seriating data, in 1957 Peter Sneath developed a method of clustering data with groupings along the sides of the matrices to show taxonomical similarities among differing strains of *E. coli*. Sneath calculated a similarity measure, S, based on a row's overlap in the matrix. Points were shaded in variations with 10% differences, and his *E. coli* data were permuted by sorting features of bacterial strains with S-values of 0–99%, choosing the topmost S-value for a given row. Then for the following rows, S-values of 0–98% were chosen, then 0–97%, and so on until all strains of *E. coli* had been grouped and printed [10]. By grouping strains based on a feature similarity, groups and subgroups can be illustrated into a dendrogram.

The addition of the dendrogram groupings along the outside of the matrix facilitates visualization of even more data by the reader. Early dendrograms were no more than groupings by section, but as time progressed, dendrograms became more complex, displaying not only groupings, but also subgroups, sub-sub-groups, and so on [2,10]. Not long after Sneath's research advancements, Robert Ling created a program to automate Hierarchical Clustering, in which many levels of subgroups are possible. Ling's method is capable of producing large and complex dendrograms along with the matrix output [11].

In the 1970s, Wilkinson developed SYSTAT, a hierarchical-based statistical display program which uses a greedy algorithm developed by Gruvaeus and Wainer [12]. However, Hierarchical Clustering with seriation does run into problems with the question of how to determine a suitable distance measure [2]. Importantly, Hierarchical Clustering does not guarantee a seriated layout. It just creates clusters and does not dictate how the leaves in the dendrogram ought to be ordered.

In 1972, William McCormick et al. released the Bond Energy Algorithm, a method for seriation based on a maximization function, assuming positive numerical data [13]. The function looks for the maximal output for all permutations of the matrix by calculating the sums of each singular datapoint, multiplied by their surrounding datapoints. Whichever permutation has the highest sum is considered the most optimal choice for seriation [14]. The Bond Energy Algorithm will be discussed at a greater length in Section 4, along with several additional popular approaches. But first we present some tips for data preprocessing.

3. Data preprocessing

It is not always the case that data obtained from a source is entirely trustworthy. There are a number of potential issues that may affect matrix data, ranging from redundant data, to inconsistencies among multiple sources, to data that are missing entirely. In this section we will discuss potential issues to watch out for in data acquisition, along with a few suggestions.

Data quality problems can stem from a number of possible sources, from single source problems with poorly designed schema or input errors, to multi source problems such as different data models among sources or even possibly contradictory data from separate sources [15].

Depending on how the data are stored, safeguards can be placed around the database. Document-oriented databases such as NoSQL databases can be extremely difficult to police given that they are ultimately key-value pairings. The pairings themselves offer flexibility such as addition of new keys/values at runtime, which can be potentially dangerous. Relational Databases tend to be significantly more rigid, yet also offer a relatively safer interaction with data input/retrieval depending on how thoroughly the schema was set up [16].

3.1 Database input errors

Regardless of how the data were obtained, it is possible, likely even, that there are input errors in the data. These errors could have occurred either through user input or an automated script reading data out of sync (possibly from an improperly encoded CSV file) or possibly data that were improperly encoded (such as '&lt' and '&gt' encodings for '<' and '>'). Thus, it is important to know the source of data. For example, it is unlikely that there will be shifted columns from a missing comma if the data were read from a database. Additionally, it is unlikely that encoding errors such as &lt would occur for non-internet scraped data. While it is unlikely, it is always wise to check for as many inconsistencies as possible.

We can protect against many of these inconsistencies in more rigid database structures such as a relational database with a schema. In a relational database with a schema, it is possible to easily protect against data input errors such as misspelled column names [15]. Note that in a NoSQL database, it is entirely likely that a new key-value pairing would be made dynamically and treated as a brand new attribute as opposed to a misspelled name of an already existing attribute [16].

Relational databases allow for schema to ensure that particular columns adhere to a specific datatype. Most modern database technologies allow for what are called stored procedures, which can allow for some level of input validation. It is not always the case that these procedures will exist, but if possible, it is always good to have some level of input validation. In relational databases, it is possible and recommended to establish foreign keys for when data are separated among tables.

3.2 Database interaction errors

If data are separated by multiple databases, ensuring consistency among records is necessary. It is likely that data will not adhere to the same schema/model across multiple databases. In cases such as this where databases do not match, it is unfortunately necessary to map each individual database schema or model to a consistent model for the matrix.

3.3 Other data errors

Not all inconsistencies will come from interactions with databases or users, and it is not always the case that they can be foreseen. In many cases, one of the largest foreseeable issues with data integrity is the case of missing data.

3.4 Data cleaning

In cleaning data, there are a number of Dos and Don'ts, however, they ultimately boil down to five steps:

- *Analysis*: Prior to knowing that the data have inconsistencies, it is necessary to manually inspect the data for inconsistencies.
- *Transformation definition*: An outlined plan for addressing inconsistencies.
- *Verification*: Ensuring that transformation methods work properly.
- *Transformation*: Any identified inconsistencies ought to be remedied.
- *Data backflow*: Depending on whether write access exists, overwriting of old, unclean data can be helpful [15].

After initial cleaning, it is likely to be the case that there exist a number of empty or null datapoints. Depending on the dataset, sometimes leaving empty data may be acceptable. In the event that leaving empty data is not acceptable, interpolation is an option.

Interpolation is a method in which some value $P(x, y)$ is derived from a series of datapoints y over some population x. There are a number of ways in which to interpolate a value, such as filling the data in with an average of all other values in the given column, randomly choosing a normally distributed value among all column values, and fitting a curve and aligning the interpolated value to the curve. Each option has its own pros and cons, however, delving into interpolation methodologies is beyond the scope of this chapter [17].

4. Algorithms

The double-seriation problem is difficult as the number of possible arrangements of rows and columns is immense and the goal for the rearrangement is not uniformly defined. For a matrix with M rows and N columns, there are $M! \times N!$ possible rearrangements. However, with the exception of biclustering [18], rearrangement of rows and rearrangement of columns may be treated as two independent processes and most approaches use the same algorithm in independent runs for each of these tasks. Given independence between permutations of rows and columns, there are $M! + N!$ possible rearrangements.

As described in Section 1, a number of approaches for the double-seriation problem have arisen over many decades. These approaches tend to vary in the precise goals they strive to attain as well as the algorithms employed for these optimizations. In this section, we provide an overview of four primary algorithms for seriation: Bond Energy Algorithm, Traveling Salesman Problem, TSP + k, and Hierarchical Clustering.

We provide a matrix of values to demonstrate the results for each of these methods. These values represent gene expression levels quantified in post mortem procedures for individuals with and without Alzheimer's disease. High expression of a gene for an individual indicates the gene was being transcribed into proteins at a greater rate than average. The motivation of this type of research is to attempt to identify molecular pathways that are increased for individuals with a disease, ultimately leading to hypothesis formation regarding suitable drug targets that may interfere with these destructive pathways. Fig. 4 depicts a small matrix of gene expression values for 10 individuals and seven genes as well as its visualization by a heatmap. In the heatmap, red color indicates high value of gene expression and blue indicates low value, as shown in the color key in Fig. 4(A).

In the following sections, we utilize gene expression data for 176 individuals with Alzheimer's disease and 188 normal control individuals, giving a total of 364 individuals. These data were collected by Amanda Myers' lab [19]. We extracted data for 100 genes that exhibited the greatest differential expression between the Alzheimer's cases and controls. The heatmap representing these data is shown in Fig. 5, where each row represents an individual and each column represents a gene. Plots for the heatmaps were generated using ClustVis [20] and Gimp. In Fig. 5C, we show the zoomed in view of the top right corner of Fig. 5A to provide a better indication of the contents of the heatmap in Fig. 5A. For simplicity, in the rest of this section, we discuss only the rearrangement of rows; column rearrangement can be applied using the same technique on a transposed matrix.

4.1 Bond Energy Algorithm

Bond Energy Algorithm (BEA) was first introduced in 1972 [13] and has been applied to diverse problems in a variety of domains. It is heavily utilized in manufacturing, in which *cell formation* is conducted to identify parts or machines with similar features or functionalities. In the manufacturing

A

0.407768	-0.05464	-0.85625	0.490587	0.360658	1.658684	0.386788	Ind1
-0.44694	-0.54461	0.196997	-0.49097	-0.17628	0.557253	-0.53705	Ind2
0.794124	-0.00609	-0.06447	0.809589	-3.0068	-0.08868	0.731862	Ind3
1.218397	0.40672	0.931902	1.228402	0.640176	0.668006	0.861576	Ind4
-1.1499	-0.50093	-4.03856	-1.36261	-0.70405	-0.81689	-0.81434	Ind5
-0.06641	-0.21975	0.171601	-0.02691	0.288927	0.064079	-0.32178	Ind6
0.418936	0.498945	0.152784	0.532225	0.385061	0.401732	0.078055	Ind7
-0.39035	-0.79363	-0.31857	-0.35644	-0.34516	0.145483	-0.51823	Ind8
1.39214	-0.28742	0.394037	0.29695	2.6003	0.814003	0.064162	Ind9
0.11496	0.656472	0.704038	0.031756	0.471152	1.57888	-0.15063	Ind10
Gene1	Gene2	Gene3	Gene4	Gene5	Gene6	Gene7	

B

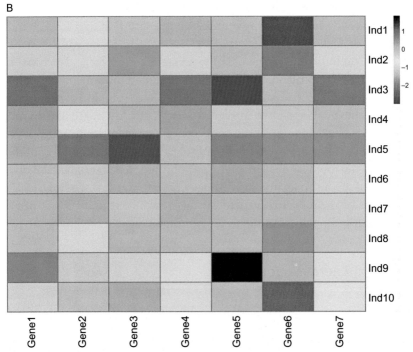

Fig. 4 (A) Gene expression levels for seven genes for 10 individuals. (B) Heatmap visually depicting the gene expression values. Red indicates high values and blue indicates low values.

Fig. 5 (A) Heatmap of unordered gene expression data. Each pixel shows the gene expression level for a specified gene in an individual. The figure on the left shows gene expression level of 100 genes (one gene per column) for 364 individuals (1 individual per row). Unit variance scaling is applied to rows. (B) Grayscale version of heatmap. (C) Zoomed in view of the top right corner of A.

domain, a comparison study between BEA, Rank Order Clustering (ROC), and Direct Clustering Analysis (DCA) was conducted by Chu and Tsai [21], revealing that in all of their trials, BEA outperformed the other methods.

BEA is also used for the *vertical segmentation* process that arises in database design [22–24], as well as path finding for autonomous vehicles [25], placement of sensors [26], and for structural improvements in software design

[27]. Notably, BEA is described in textbooks on database design [28] and data mining [29]. Two biological applications include the use of BEA for the seriation of heatmaps containing gene expression data [30] and to identify clusters of interacting proteins [31].

BEA aims to maximize the *bonds* between rows of the matrix, based on an intuition that high values will attract to other high values. More specifically, a measure of effectiveness (ME) is introduced and defined as follows for a nonnegative $M \times N$ matrix A with entries a_{ij}:

$$\max ME = \sum_{i=1}^{M} \sum_{j=1}^{N} a_{ij}\left(a_{i+1,j} + a_{i-1,j}\right) \tag{1}$$

In order to accommodate the first and last rows of the matrix, two additional rows are defined as the zero and $M+1$ rows where all entries in the new rows are zeros. Consequently, the elements in the first row are multiplied by only the entries in the second row; the elements in the last row are multiplied by only the elements in the $M-1$ row.

The objective of BEA is to maximize ME. The basic intuition is that maximization of this function will tend to place rows with large values next to rows with large values for the same columns. This problem can be rewritten as a Quadratic Assignment Problem (QAP) [13]. QAP is a combinatorial problem that commonly arises in various forms in the field of operations research [32]. A basic formulation of the problem follows in which the locations for facilities, such as factories, are determined in order to reduce the amount of transportation (flow) required between the facilities. Given a set of M facilities and M locations (same number of facilities and locations), along with distances between each pair of locations and flow between each pair of facilities, assign each facility to a unique location such that the sum of the distances multiplied by flows is minimized. This problem is NP-hard [33].

Due to computational demands for identifying optimal solutions to the BEA objective (Eq. 1), algorithms that approximately optimize ME have been developed. McCormick et al. introduced a simple greedy strategy when they defined the ME function [13] and this method quickly became popular in diverse domains, as previously described. First, a row is randomly chosen and placed in the new matrix. Second, the increase in the ME value is computed for the placement of each unused row before and after the previously placed rows in the new matrix. The row with the maximum contribution is selected and placed in the new matrix. The second step is repeated until all rows are placed.

It should be noted that BEA does not allow negative values in the matrix. When negative values are present, this requirement can easily be managed by shifting values. Each value in the matrix can be increased by $|x|$, where x is the smallest value, resulting in a nonnegative matrix to which BEA can be applied. After the rearrangement, all values should be decreased by $|x|$ to return to the original values.

We previously reported two pitfalls for BEA analyses [5]. The first of these arises for binary and ternary matrices, as well as other matrices with large numbers of zero entries. This pitfall becomes irrelevant in the context of interest as heatmaps typically do not have this property. However, the second pitfall is highly relevant for heatmaps. BEA moves high values closer together without regard for low values. Consequently, clusters of low values may be split apart. This is serious in the domain of gene expression as low-valued clusters may be of primary interest. For example, for gene expression heatmaps, low-valued clusters might represent a biological pathway that is being repressed or a large protein complex that is not being produced at normal levels.

Note that even if optimal solutions to the NP-hard objective were found, this pitfall would still be of grave concern. As shown in Eq. 1, each of the entries except those in the first and last rows contribute to the summation three times: when in the current row, the row above the current row, and the row below the current row. In contrast, the first and last rows only contribute twice. Entries in the first row do not contribute when serving as the row below the current row, and entries in the last row do not contribute as the row above the current row. Consequently, rows with the lowest values will likely appear in the first and last rows. Rows with similarly low values will be adjacent and the highest-valued rows will tend to cluster near the center of the matrix, splitting the low-valued rows into two separated portions. Fig. 6 shows the rearrangement of the Alzheimer's study data in Fig. 5 using BEA.

Complexity. Optimally solving BEA is NP-hard. Solving it using McCormick et al.'s approximation approach requires $O(M^2N)$ time, where M is the number of rows and N is the number of columns. Space requirements for this algorithm are nominal as it is possible to solve in place and store row indices in an $M \times 1$ array.

4.2 Traveling Salesman Problem Model

In 1974, Jan Karel Lenstra observed that data seriation could be cast as a Traveling Salesman Problem [34,35]. The Traveling Salesman Problem

A   B

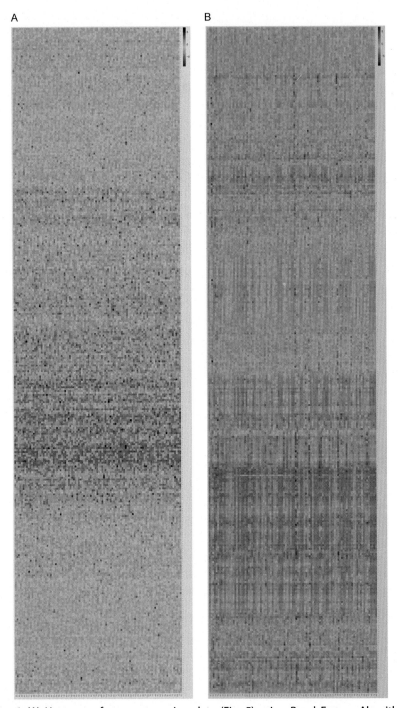

Fig. 6 (A) Heatmap of gene expression data (Fig. 5) using Bond Energy Algorithm. (B) Grayscale version of heatmap.

(TSP) is a well-studied combinatorial problem with many diverse applications. The TSP has a long and rich history. It was included in a traveling salesmen "how to" book in 1832 [36] and has been studied by mathematicians at least since the 1920s [37]. Being an archetypal NP-hard problem, research ventures have stimulated the development of several general-purpose solvers including Branch-and-Bound, Branch-and-Cut, Cut-and-Solve, Simulated Annealing, and Lagrangian Relaxations [38–45].

TSP attempts to minimize the cost of travel for a salesman needing to visit a set of cities and return to the starting city. Given this set of cities denoted as the vertices V in the graph, and the cost of travel between each pair of cities as the weight of edges between vertices, the TSP is to find the route for which the sum of the costs of travel is minimized. TSP also ensures that none of the cities, except the first, is visited more than once. More formally:

$$\min Z = \sum_{i \in V} \sum_{j \in V} x_{ij} c_{ij} \tag{2}$$

such that:

$$\sum_{i \in V} x_{ij} = 1, \quad \forall j \in V \tag{3}$$

$$\sum_{j \in V} x_{ij} = 1, \quad \forall i \in V \tag{4}$$

$$\sum_{i = W} \sum_{j = W} x_{ij} < |W|, \quad \forall W \subset V, \ W \neq \varnothing \tag{5}$$

$$x_{ij} \in \{0, 1\}, \quad \forall i, j \in V \tag{6}$$

where c_{ij} is equal to the cost of travel between cities i and j, and x_{ij} are binary decision variables. When city j is visited directly after city i, x_{ij} is set equal to one, otherwise it is set to zero. Consequently, the objective function (Eq. 2) will yield the cost of travel for the tour, which is to be minimized as indicated. Constraints (Eq. 3) require that each city is entered exactly once and constraints (Eq. 4) require that each city is exited from exactly once. Constraints (Eq. 5) are called the "subtour elimination" constraints. Without these constraints, an optimal solution could contain any number of cycles. When a cluster of cities are close together, they tend to form their own independent cycle thereby omitting relatively long distances to connect it with other clusters. The subtour elimination constraints require that every proper subset, W, of the cities have fewer edges with their x_{ij} values set to one than the cardinality of the subset. Because a cycle contains the same number of cities as traversed edges, these constraints ensure that no cycles can be formed. Note that these constraints hold for proper subsets and not when W is equal to V, as the number of edges with x_{ij} equal to one in the optimal solution needs to equal to the number of cities, thereby providing the requisite cycle to start and finish at the home city. Constraints (Eq. 6) require that the decision variables, x_{ij}, are either zero or one and are not allowed to have fractional values (either an edge is traversed or it is not).

Each set of constraints (Eqs. 3 and 4) are comprised of $|V|$ constraints. How many subtour elimination constraints (Eq. 5) are there? To answer this, we need to determine the number of subsets possible. Since each city could either be in a given subset or not, we have $2^{|V|} - 1$ proper subsets (we omit the subset containing all of the cities). An exponential number of constraints is not at all practical—the common approach is to ignore these constraints, solve the TSP, and check for subtours. If any subtours arise, a specific constraint omitting one of them is added to the problem formulation and the problem is solved again. This process is repeated until a TSP solution is found without any subtours.

How did Lenstra cast the seriation problem as a TSP? First, each row is treated as a city and similarities or differences between rows of the matrix are computed using a metric of choice. When similarities, such as Pearsons Correlation Coefficient [46] are computed, the values are inverted to represent differences, or distances, between the rows. These distances are used for the cost matrix values c_{ij}. Solving the TSP with these values produces a linear ordering of the rows such that the sum of the differences between each pair of adjacent rows is minimized. However, this sum includes the distance between the last row and the first row as the TSP makes a full cycle, returning to the home city. Lenstra simply added a *dummy* city to the problem definition to address this issue. The cost between the dummy city and all other cities is set to some constant value, C. After solving this revised problem, the dummy city is used to mark the separation between the last and first rows and form a linear path, thereby providing the rearrangement for the matrix. For the optimal solution to this revised problem, we previously proved that the actual distance between the two cities separated by the dummy city is greater than or equal to any of the distances between any pairs of adjacent cities in the tour, and that the total distance of the TSP path is the smallest possible [5].

The objective for seriation using Lenstra's TSP follows:

$$\min TSP = \sum_{i=1}^{M-1} c(i, i + 1) \tag{7}$$

where $c\,(i,j)$ is the dissimilarity between rows i and j.

The objective for Bond Energy Algorithm (Eq. 1) is different from the objective for Lenstra's TSP approach (Eq. 7). In the former, the adjacent row values are multiplied together and the sum of these products is maximized thereby leading to the congregation of high values, without regard for low values, which tend to be split apart and placed at the top and bottom of the matrix. For the latter, the sum of the dissimilarities between each pair of adjacent rows is minimized, thereby attracting similar rows together, regardless of whether they are comprised of high, low, or moderate values. TSP also easily handles rows containing any type of mixture of high, low, and moderate values. This is especially important for heatmap seriation as such mixtures are common.

It is well known that optimally solving the TSP is NP-hard [47]. The number of possible tours for $M+1$ cities is $M!$. To provide some context, for $M = 100$, there are 5050 decision variables and roughly 10^{158} feasible rearrangements. Note that it is estimated that there are approximately 10^{80} subatomic particles in the known universe.

At first glance, it may seem that optimally solving TSP for practical problems is computationally infeasible, but due to vast research into this domain, there have been many advances in the field to facilitate great success for some types of TSP instances. Notably, an 85,900-city TSP was solved to optimality in 2006 [48]. How could such a feat be accomplished? The researchers used a strategy called Branch-and-Cut to prune away most of the solution space without risk of pruning away the optimal solution [49]. The basic idea underlying Branch-and-Cut is that the solution space is partitioned into a search tree in which a pair of child nodes are generated from a parent node by setting an x_{ij} decision variable to either zero or one. Lower and upper bounds on the optimal solution are computed as the tree grows and when the lower bound on a node is greater than or equal to the upper bound, the entire subtree rooted at that node can be pruned without loss of optimality.

Much of the TSP research has focused on identifying methods to tighten lower bounds for instances in which cities are represented by coordinates in a two-dimensional plane and the Euclidean distance is used to generate the cost matrix. For these instances, distances are symmetric, and the triangle inequality[a] holds. We expect correlations between rows of a heatmap to be symmetric, but the triangle inequality might not hold. In general, state-of-the-art open-source solvers, such as Concorde (http://www.

---

[a] Triangle inequality states that the sum of distances from node A to node B and from node B to node C is greater than or equal to the direct distance from node A to node C, if it exists.

math.uwaterloo.ca/tsp/concorde/index.html), and commercial solvers, such as Gurobi (https://www.solver.com/gurobi-solver-engine) and IBM's Cplex (https://www.ibm.com/analytics/cplex-optimizer), have mixed results when solving TSP instances, even when both properties are present, but usually are capable of solving rearrangements for moderate sized heatmaps. It should be noted that both Cplex and Gurobi currently offer free access for academic institutions and non-profit research.

Optimally solving larger heatmaps might be possible by highly parallelizing the search. Unfortunately, due to the inherently sequential nature of tightening lower bounds, large-scale parallelization of Branch-and-Cut search has not been effective. On the other hand, Branch-and-Bound [50,51], and Cut-and-Solve [38], are both easily parallelized. However, parallelized Branch-and-Bound results still tend to fall short of Branch-and-Cut performance and parallelized software for Cut-and-Solve is not currently publicly available.

For these reasons, TSP instances that remain intractable for current optimal solvers are typically solved using approximation methods. Such solvers are numerous and widely varied, having arisen from virtually every domain in Artificial Intelligence and Operations Research [48,52]. Some are extremely fast and can handle millions of cities. Performances are mixed across the techniques, and there is generally a trade-off between speed and quality, but this is not consistent. Indeed, one algorithm may be fast and accurate for one instance and perform poorly for a similar instance. Some of these methods, such as our own [38], are *anytime* solvers, in that they can be stopped at any time and the best result found so far is returned, but if left to run long enough, will eventually produce the optimal solution. Fig. 7 shows the rearrangement of our example problem (gene expression data in Fig. 5) using an optimal TSP solution.

Complexity. Optimally solving the TSP is NP-hard. Furthermore, some strategies, such as Branch-and-Cut, are memory intensive and memory exhaustion may arise within a few hours of computation time. Approximate solutions are widely varied in computational complexity and memory usage.

4.3 TSP + *k*

During our previous study of heatmaps for gene expression data, we inadvertently reinvented Lenstra's TSP solution. In the domain of gene expression studies, and indeed in most applications of heatmaps, we expect to see *cluster* properties, in which items within a given cluster show high similarities

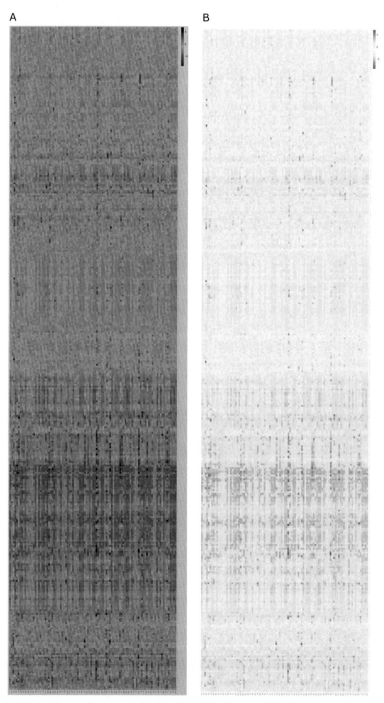

Fig. 7 (A) Heatmap of gene expression data (Fig. 5) using TSP. (B) Grayscale version of heatmap.

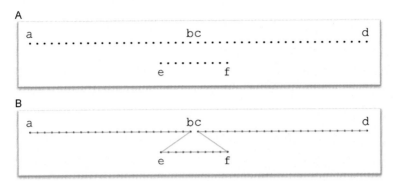

Fig. 8 Pitfall for TSP clustering. Assuming Euclidean distance, (A) represents two linear clusters of points, the first from a to d, and the second from e to f. The optimal TSP clustering is shown in (B). The two clusters are joined into a linear ordering by traversing from b to e and f to c, as this minimizes the overall summation. Although b and c are very close, they are separated by 10 nodes in the linear ordering.

to each other and clusters may be quite distinct from each other. We quickly realized a major pitfall for using Lenstra's TSP for rearranging data that tends to fall into natural clusters [5]. Inter-cluster distances between clusters tend to be larger than intra-cluster distances between objects co-existing together in respective clusters. Consequently, inter-cluster distances tend to dominate the summation in Eq. 7 and the rearrangement may be skewed in order to minimize these large inter-cluster distances. Fig. 8 shows a simple toy example of this pitfall.

We previously presented a solution to address this pitfall [5,53] and named it TSP + k for reasons that will become apparent shortly. Next, we present our revised objective function, then we describe a simple technique to optimize this function.

The TSP + k objective function follows:

$$\min TSP(k) = \sum_{i=1}^{k} \sum_{j=u_i}^{v_i-1} c(j, j+1) \tag{8}$$

where u_i and v_i represent the starting and ending rows for cluster i. The inner summation adds up the distances between rows within a given cluster, i, and the outer loop sums up these values for k clusters. Thus, the intra-cluster distances are included while the inter-cluster distances are omitted. In other words, the minimization will only be performed over the intra-cluster edges and the distances between clusters are completely disregarded. This strategy simultaneously identifies the optimal cluster memberships and the ordering of the rows within each cluster. Distances between clusters are allowed to be arbitrarily large.

A simple technique to optimize Eq. (8) is to add k dummy cities to the TSP model of the problem instance, where k is the number of desired clusters. After solving the TSP instance, the dummy cities are removed and their locations indicate cluster boundaries. In this way, the dummy cities divide up the TSP path into k discrete paths. In addition to resolving the TSP pitfall, this approach offers two additional benefits. First, granularity can be defined by choosing a suitable range of values for k. In some cases, it is desirable to identify only a few clusters while in others, a higher granularity may be desired. Second, the cluster boundaries are clearly defined by the TSP solution. Various methods have been employed to discern cluster boundaries for alternative seriation methods, such as visual inspection and computational strategies (e.g. [31,54]), but TSP + k provides the optimal cluster boundaries automatically. Fig. 9 shows the rearrangement of our example problem (gene expression data in Fig. 5) using TSP + k.

Complexity. Optimally solving TSP + k has the same complexity as TSP and is NP-hard. There are k more cities added to the model, but this number tends to be small in comparison with the number of rows. TSP + k can also be solved using standard TSP approximation algorithms with similar overall complexity.

4.4 Hierarchical Clustering

Perhaps the most commonly-used algorithm for heatmap seriation is Hierarchical Clustering. Hierarchical Clustering is a general clustering method used for detecting a hierarchy of communities of nodes in network models. Importantly, it does not enforce any type of linear ordering within the clusters. For heatmaps, each row is modeled as a node in a network and similarities/distances between each pair of rows is computed and modeled as edges between the relevant nodes in the network. In contrast to the other methods discussed in this chapter, Hierarchical Clustering first identifies clusters within this network model and then flattens these clusters into a linear ordering on the matrix.

Hierarchical Clustering is implemented in a variety of ways, dependent upon several options. First, the general strategy can be agglomerative, in which each node begins as a cluster and clusters are iteratively joined; or

A   B

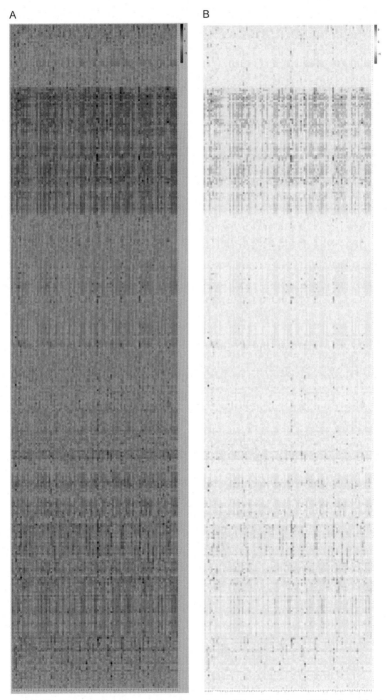

Fig. 9 (A) Heatmap of gene expression data (Fig. 5) using TSP + k with $k = 4$. (B) Grayscale version of heatmap.

divisive, in which all nodes are initially placed in a single cluster and clusters are iteratively divided, typically by dividing the cluster that currently holds the most nodes. Second, the linkage/splitting criteria can vary. Some examples for agglomerative clustering include *single-linkage*, for which the two clusters that have the minimum distance between their two closest nodes are chosen to be joined, and *centroid linkage*, which chooses the two clusters with the minimum distance between their respective centroids. Divisive clustering uses a variety of splitting strategies, such as K-Means Clustering. Third, as in TSP and TSP + k, the similarity/distance measure may take many forms, based on metrics such as Euclidean Distance, Hamming Distance, or Pearson's Correlation Coefficient.

As in TSP + k, Hierarchical Clustering holds the advantage of being able to choose the granularity of clustering. The results of each step of the algorithm are retained in a dendrogram and researchers may examine this dendrogram, along with the heatmap, to determine a level of clustering granularity that best meets the needs of the current study.

Although Hierarchical Clustering defines the clusters in which nodes reside, it does not provide a deterministic method for the rearrangement of the corresponding rows. Within each cluster the rows can be ordered in various ways, resulting in 2^{M-1} unique layouts for a given clustering result [54]. Consequently, some layouts might place clusters in between two clusters that are similar to each other and visual inspection to determine granularity might miss this similarity.

In general, optimally solving either agglomerative or divisive Hierarchical Clustering requires exhaustive $O(2^M)$ computation time. Most implementations identify approximate solutions using a greedy strategy, in which, for each step, the *best* solution for the current state is chosen and committed, without any reevaluations at later stages. Agglomerative approaches determine the two closest clusters at the current time and join them to produce a new cluster. Divisive approaches use some strategy, such as k-means clustering, to determine how to split the chosen cluster.

It is important to note that both of these approximation techniques may present bias toward spherical clusters. k-means clustering, as well as popular methods for agglomerative clustering, utilize distances to centroids of clusters or distances between the two most extreme points for two clusters. This practice tends to produce spherical clusters. Elongated and other cluster configurations may be split apart due to some of the cluster nodes being closer to

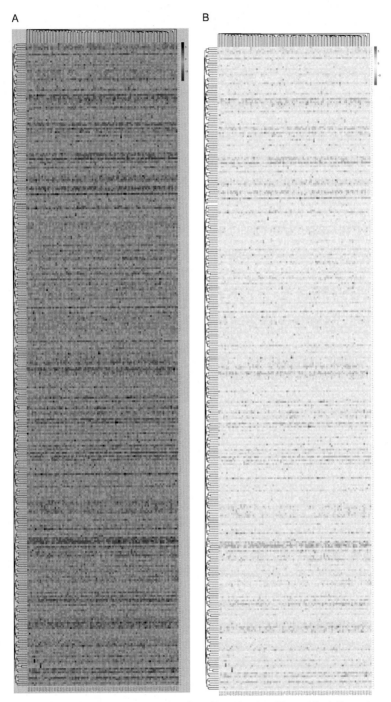

Fig. 10 (A) Heatmap of gene expression data (Fig. 5) using Hierarchical Clustering with default settings provided in ClustVis [20]. (B) Grayscale version of heatmap.

another cluster's centroid than its own or due to a large *diameter*. This may be problematic in some domains, such as genetics, in which clusters may have intricate structures.

There are alternative general-purpose clustering methods that may not enforce sphericity, but they generally are not applied to heatmaps. Consider for example, Newman and Girvan's popular modularity objective function [55], which does not enforce sphericity and rather aims to group objects with many pairwise similarities together and separates clusters with few shared similarities. Optimizing this objective is NP-hard, but approximation algorithms are prevalent in the literature and range from extremely fast [56] to slow, but exceptionally accurate [57]. Most of these algorithms are designed for networks comprised of nodes and unweighted edges. A heatmap seriation problem can be modeled this way by representing each row as a node, determining a significant similarity score, and placing an edge between each pair of nodes representing two rows with significant similarity. On the other hand, when an algorithm utilizes weights for the edges, the edges may include the value of the similarity score.

Why are not these alternatives commonly employed? Most algorithms, including those designed for modularity, produce a clustering result without information for different levels of granularity. When forced to place the rows into a heatmap, the ordering is completely arbitrary within each cluster and the clusters are arbitrarily placed. Although Hierarchical Clustering has 2^{M-1} unique layouts possible, the chosen layout, along with its dendrogram, provides more insightful information for visual inspections. Fig. 10 shows the rearrangement of our example problem (gene expression data in Fig. 5) using Hierarchical Clustering.

Complexity. Optimally solving Hierarchical Clustering requires $O(2^M)$ computation time. Typical approximate solutions for agglomerative clustering require $O(M^3)$ computation time and $O(M^2)$ memory. Use of a heap can reduce computation time to $O(M^2 \log M)$, but increase memory requirements. Approximate solutions to divisive clustering require various resources based on the approach used to determine how to split the clusters.

4.5 Algorithms summary

Table 1 provides a brief summary of the pros and cons for each algorithm when solving optimally or approximately.

Table 1 Summary of four seriation methods.

Algorithm	Optimal solutions			Approximate solutions		
	Complexity	Pros	Cons	Complexity	Pros	Cons
Bond Energy Algorithm	NP-hard	Brings large values together	Tends to split clusters with low values apart, which may be especially detrimental for heatmap seriation	$O(M^2N)$	Has been shown to provide desirable results in multiple domains	Tends to split clusters with low values apart, which may be especially detrimental for heatmap seriation
Traveling Salesman Problem	NP-hard	Rearranges rows such that sum of differences between rows is absolutely minimized; robust for rows containing high, moderate, and low values as well as any mixture of values	Natural clusters may be split in order to minimize jump to adjacent clusters	Widely varied	Able to solve very large heatmaps; anytime algorithms can directly facilitate capturing promising solutions for given computational resources	Not readily apparent which algorithm will provide the best solution for given computational resources
TSP + k	NP-hard	Rearranges rows such that sum of differences between rows within clusters is absolutely minimized while ignoring differences between clusters	Provides optimal linear ordering (for visualization), but this ordering might not provide optimal clustering when linear ordering is not needed (e.g. for biological relevance)	Widely varied	Able to solve very large heatmaps; anytime algorithms can directly facilitate capturing promising solutions for given computational resources	Not readily apparent which algorithm will provide the best solution for given computational resources

| Hierarchical Clustering | $O(2^M)$ | Provides dendrogram allowing visualization of various clustering granularities | Linear ordering is not optimal and dendrogram can be ordered in 2^{M-1} unique layouts | $O(M^3)$ for typical agglomerative; varied for divisive | Numerous open-source software packages available, most will plot dendrogram with heatmap | For most available software packages, clustering is biased toward increased sphericity regardless of true cluster configuration |

Complexities are shown in terms of the number of rows, M. Double seriation would require repeating the method on the transposed matrix so overall complexity should be based on the maximum of M and N for the original matrix.

5. Summary

Heatmaps provide insightful information in a comprehensive manner. They have been utilized for decades, yet the approaches for rearranging the rows and columns are often applied without full realization of underlying assumptions and properties of the method.

In this chapter, we highlighted some of the characteristics of optimal and approximate computations for four rearrangement clustering approaches: Bond Energy Algorithm, Traveling Salesman Problem Model, TSP + k, and Hierarchical Clustering. Optimally solving any of these models is computationally demanding, yet provides some benefits over approximate solutions when adequate computational resources are available.

Bond Energy Algorithm aims to bring large values together and has been shown to provide desirable results in multiple domains. However, it tends to split clusters with low values apart, pushing them to the top and bottom of the heatmap. This property can be especially detrimental for heatmap construction.

The Traveling Salesman Problem Model minimizes the sum of the differences between adjacent rows. It is robust for rows containing high, moderate, and low values, as well as mixtures of values. There exists a plethora of approximation algorithms, providing a wide range of computation requirements and solution qualities. In fact, there are implementations that can solve very large heatmaps with modest computation. Furthermore, anytime solvers can compute until resources are exhausted and provide the best approximate solution found or the optimal solution if time is adequate. The major drawback for this approach is that natural clusters may be split in order to minimize the jumps to adjacent clusters.

TSP + k addresses the pitfall of the Traveling Salesman Problem Model by adding dummy cities to allow arbitrarily large jumps between natural clusters. After the additions, the instance can be solved by any Traveling Salesman Problem solver, thereby providing the benefits of the extensive research in this field and the vast array of implemented solvers. Although solving this problem optimally provides an exemplary linear ordering for heatmap seriation, this linear ordering might not be needed for actual clustering of data—it is primarily of importance for visualization.

Finally, Hierarchical Clustering is a greedy approach that provides a dendrogram, thereby allowing visualization of various clustering granularities. There are numerous open-source software packages available and most will plot the dendrogram with the heatmap, providing valuable information. However, the linear ordering for Hierarchical Clustering is not optimal and the dendrogram can be ordered in 2^{M-1} unique layouts for M rows (columns). Also, most software available induce a bias toward increased sphericity, regardless of true cluster configuration.

The use of heatmaps for data visualization has a long and rich history. Yet progress continues to be made in methods to properly preprocess data and the development of algorithms to yield greater insights into patterns hidden within the data. Furthermore, current state-of-the-art approaches bear tradeoffs in performance and it is clear that this field continues to hold potential for innovative improvements that may provide greater opportunities for scientific discovery via visualization.

5.1 Future directions

One movement that has yet to be adopted by the field entirely, but has been growing, is the adoption of color-blind palettes. Red-green color blindness, Deuteranopia, is the most common form of this trait. Color blindness is a prevalent trait affecting a significant portion of the population. Being an X-linked recessive trait, it affects males more commonly than females. For example, individuals with European ancestry exhibit rates of Deuteranopia around 8% for males and 0.5% for females [58]. Despite the prevalence of this trait, huge numbers of heatmaps are generated in which red and green represent high and low values, respectively. Fig. 11 illustrates a red-green heatmap and the same heatmap as observed by an individual with Deuteranopia. ClustVis uses a palette that is more compliant for these individuals, resulting with distinguishable heatmaps, as shown in Fig. 12. As of now, color-blind options for data visualization are not ubiquitous, creating a great disadvantage for many researchers.

Another pressing issue is the need for increased user-friendliness for existing approaches. Such tools are currently most readily available for Hierarchical Clustering (e.g. ClustVis [20]) while other methods typically require significant effort by the user. To that end, we have developed open-source code to streamline clustering using TSP + k. Note that this software can also perform TSP clustering by setting k equal to one. This package is available at: https://github.com/DaTino/tspk.

Finally, the amount of data being generated is rapidly accelerating, leading to barriers for future research. Many of the current methods are likely to run into an issue of either prohibitively long execution times or exhausting

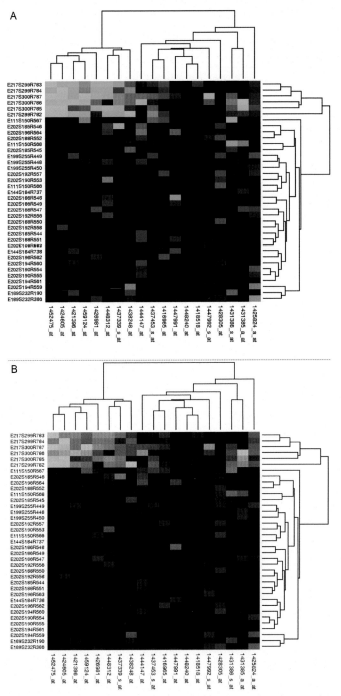

Fig. 11 (A) A red-green heatmap, for which red indicates high values and green indicates low values (https://commons.wikimedia.org/wiki/File:Heatmap.png). (B) The same heatmap as observed by individuals with Deuteranopia, the most common form of color blindness. Individuals with deuteranopia lack the cone cells that respond to the color green and are unable to distinguish red from green. Note that the red cells, which indicate high values, appear similar to the green cells, which indicate low values. Image generated using GIMP.

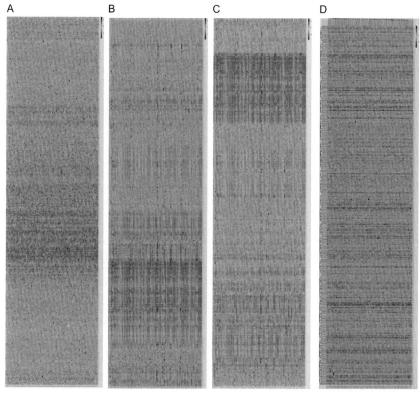

Fig. 12 Perception of ClustVis heatmaps by individuals with Deuteranopia for matrices seriated using (A) Bond Energy Algorithm (Fig. 6), (B) Traveling Salesman Problem (Fig. 7), (C) TSP + k (Fig. 9), and (D) Hierarchical Clustering (Fig. 10).

memory. It seems reasonable to predict that some major leaps for heatmap generation, and matrix rearrangement/manipulation in general, will come from using high-performance computing. Such approaches might include the use of GPUs, quantum computing, and/or parallel processing.

Another issue with increasing data size is the visualization of massive heatmaps. A promising direction might be the development of tools capable of nimbly zooming in and out of matrices while providing relevant information for the observed cells. In a related vein, the future may present ingenious methods that facilitate the insights provided by examining heatmaps exhibiting natural clusters without imposing the linearity constraints invoked by current seriation methods.

References

[1] D. Adam, Metric prefixes sought for extreme numbers, Science (New York, NY) 363 (6428) (2019) 681. https://doi.org/10.1126/science.363.6428.681.

[2] L. Wilkinson, M. Friendly, The history of the cluster heat map, Am. Stat. 63 (2) (2009) 179–184. https://doi.org/10.1198/tas.2009.0033.

[3] P.A. Kidwell, A history of mathematical tables: from Sumer to spreadsheets (review), Technol. Cult. 45 (3) (2004) 662–664. https://doi.org/10.1353/tech.2004.0136.

[4] T. Loua, Atlas Statistique de la Population de Paris, J. Dejey & cie, Paris, 1873. Retrieved from https://www.worldcat.org/title/atlas-statistique-de-la-population-de-paris/oclc/690655859.

[5] S. Climer, W. Zhang, Rearrangement clustering: pitfalls, remedies, and applications, J. Mach. Learn. Res. 7 (2006) 919–943. Retrieved from http://jmlr.csail.mit.edu/papers/v7/climer06a.html.

[6] I. Liiv, Seriation and matrix reordering methods: an historical overview, Statist. Anal. Data Min. 3 (2) (2010) 70–91. https://doi.org/10.1002/sam.10071.

[7] W.M.F. Petrie, Sequences in prehistoric remains, J. Anthropol. Inst. G. B. Irel. 29 (3/4) (1899) 295. https://doi.org/10.2307/2843012.

[7a] R.M. Gray, Toeplitz and circulant matrices: a review, Found. Trends Commun. Inf. Theory 2 (3) (2005) 155–239. https://doi.org/10.1561/0100000006.

[8] A. Soltysiak, P. Jaskulski, Czekanowski's diagram a method of multidimensional clustering, in: CAA1998. New Techniques for Old Times. Computer Applications and Quantitative Methods in Archaeology. Proceedings of the 26th Conference, Barcelona, March 1998 (BAR International Series 757), Barcelona, Spain, 1999, pp. 175–183. Retrieved from https://www.academia.edu/9281897/Czekanowskis_diagram._A_method_of_multidimensional_clustering.

[9] L. Wilkinson, M. Friendly, History corner the history of the cluster heat map, Am. Stat. 63 (2) (2009) 179–184. https://doi.org/10.1198/tas.2009.0033.

[10] P.H. Sneath, The application of computers to taxonomy, J. Gen. Microbiol. 17 (1) (1957) 201–226. Retrieved from http://www.ncbi.nlm.nih.gov/pubmed/13475686.

[11] R.L. Ling, A computer generated aid for cluster analysis, Commun. ACM 16 (6) (1973) 355–361. https://doi.org/10.1145/362248.362263.

[12] L. Wilkinson, SYSTAT, Wiley Interdiscip. Rev. Comput. Stat. 2 (2) (2010) 256–257. https://doi.org/10.1002/wics.66.

[13] W.T. McCormick, P.J. Schweitzer, T.W. White, Problem decomposition and data reorganization by a clustering technique, Oper. Res. 20 (5) (1972) 993–1009. https://doi.org/10.1287/opre.20.5.993.

[14] P. Arabie, L.J. Hubert, The bond energy algorithm revisited, IEEE Trans. Syst. Man Cybern. 20 (1) (1990) 268–274. https://doi.org/10.1109/21.47829.

[15] E. Rahm, H.H. Do, Data cleaning: problems and current approaches, Informatica 23 (4) (2000) 3–13.

[16] R. Cattell, Rick, Scalable SQL and NoSQL data stores, ACM SIGMOD Rec. 39 (4) (2011) 12. https://doi.org/10.1145/1978915.1978919.

[17] D. Shepard, Donald, A two-dimensional interpolation function for irregularly-spaced data, in: Proceedings of the 1968 23rd ACM National Conference on ACM '68, ACM Press, New York, NY, 1968, pp. 517–524. https://doi.org/10.1145/800186.810616.

[18] S.C. Madeira, A.L. Oliveira, Biclustering algorithms for biological data analysis: a survey, IEEE/ACM Trans. Comput. Biol. Bioinform. 1 (1) (2004) 24–45. https://doi.org/10.1109/TCBB.2004.2.

[19] J.A. Webster, J.R. Gibbs, J. Clarke, M. Ray, W. Zhang, P. Holmans, …D.S. McCorquodale, Genetic control of human brain transcript expression in Alzheimer disease, Am. J. Hum. Genet. 84 (4) (2009) 445–458. https://doi.org/10.1016/j.ajhg.2009.03.011.

[20] T. Metsalu, J. Vilo, ClustVis: a web tool for visualizing clustering of multivariate data using principal component analysis and heatmap, Nucleic Acids Res. 43 (W1) (2015) W566–W570. https://doi.org/10.1093/nar/gkv468.

[21] C.H. Chu, M. Tsai, A comparison of three array-based clustering techniques for manufacturing cell formation, Int. J. Prod. Res. 28 (8) (1990) 1417–1433. https://doi.org/10.1080/00207549008942802.

[22] J.A. Hoffer, D.G. Severance, The use of cluster analysis in physical data base design, in: Proceedings of the 1st International Conference on Very Large Data Bases— VLDB '75, ACM Press, New York, NYs, 1975, p. 69. https://doi.org/10.1145/1282480.1282486.

[23] S. Navathe, S. Ceri, G. Wiederhold, J. Dou, Vertical partitioning algorithms for database design, ACM Trans. Database Syst. 9 (4) (1984) 680–710. https://doi.org/10.1145/1994.2209.

[24] H. Rahimi, F.-A. Parand, D. Riahi, Hierarchical simultaneous vertical fragmentation and allocation using modified bond energy algorithm in distributed databases, Appl. Comput. Inform. 14 (2) (2018) 127–133. https://doi.org/10.1016/J.ACI.2015.03.001.

[25] Y.-h. Chang, B. Hyun, A.R. Girard, Path planning for information collection tasks using bond-energy algorithm, in: 2012 American Control Conference (ACC), IEEE, 2012, pp. 703–708. https://doi.org/10.1109/ACC.2012.6314671.

[26] W. Lu, R. Wen, J. Teng, X. Li, C. Li, Data correlation analysis for optimal sensor placement using a bond energy algorithm, Measurement 91 (2016) 509–518. https://doi.org/10.1016/J.MEASUREMENT.2016.05.089.

[27] N. Gorla, K. Zhang, Deriving program physical structures using bond energy algorithm, in: Proceedings—6th Asia Pacific Software Engineering Conference, APSEC 1999, IEEE Computer Society, 1999, pp. 359–366. https://doi.org/10.1109/APSEC.1999.809624.

[28] M.T. Özsu, P. Valduriez, Principles of Distributed Database Systems, third ed., Springer New York, New York, NY, 2011. https://doi.org/10.1007/978-1-4419-8834-8.

[29] M.H. Dunham, Data Mining: Introductory and Advanced Topics, Pearson, 2003.

[30] Y. Liu, B. Ciliax, A. Pivoshenko, J. Civera, V. Dasigi, A. Ram, … S. Navathe, Evaluation of a new algorithm for keyword-based functional clustering of genes, in: 8th International Conference on Research in Computational Molecular Biology (RECOMB-04) (p. poster), 2004.

[31] R.L.A. Watanabe, E. Morett, E.E. Vallejo, Inferring modules of functionally interacting proteins using the bond energy algorithm, BMC Bioinform. 9 (1) (2008) 285. https://doi.org/10.1186/1471-2105-9-285.

[32] E.L. Lawler, The quadratic assignment problem, Manag. Sci. 9 (4) (1963) 586–599. INFORMS. https://doi.org/10.2307/2627364.

[33] M.R. Garey, D.S. Johnson, Computers and Intractability: A Guide to the Theory of NP-Completeness, W.H. Freeman, 1979. Retrieved from https://dl.acm.org/citation.cfm?id=574848.

[34] J.K. Lenstra, Clustering a data array and the traveling-salesman problem, Oper. Res. 22 (2) (1974) 413–414. https://doi.org/10.1287/opre.22.2.413.

[35] J.K. Lenstra, A.H.G.R. Kan, Some simple applications of the travelling salesman problem, Oper. Res. Q. (1970–1977) 26 (4) (1975) 717. https://doi.org/10.2307/3008306.

[36] B.F. Voigt, Der Handlungsreisende—wie er sein soll und was er zu thun hat, um Auftraege zu erhalten und eines gluecklichen Erfolgs in seinen Geschaeften gewiss zu sein—Von einem alten Commis-Voyageur, Voigt, Ilmenau, 1832. Retrieved from https://www.worldcat.org/title/handlungsreisende-wie-er-sein-soll-und-was-er-zu-thun-hat-um-auftrage-zu-erhalten-und-eines-glucklichen-erfolgs-in-seinen-geschaften-gewi-zu-sein-mit-e-titelkupf/oclc/258013333.

[37] D. Applegate, R. Bixby, V. Chvatal, B. Cook, Finding cuts in the TSP (A preliminary report), in: Center for Discrete Mathematics & Theoretical Computer Science, 1995 Retrieved from https://dl.acm.org/citation.cfm?id=868329.

[38] S. Climer, W. Zhang, Cut-and-solve: an iterative search strategy for combinatorial optimization problems, Artif. Intell. 170 (8–9) (2006) 714–738. https://doi.org/10.1016/j.artint.2006.02.005.

[39] G.A. Croes, A method for solving traveling-salesman problems, Oper. Res. 6 (6) (1958) 791–812. https://doi.org/10.1287/opre.6.6.791.

[40] G. Dantzig, R. Fulkerson, S. Johnson, Solution of a large-scale traveling-salesman problem, J. Oper. Res. Soc. Am. 2 (4) (1954) 393–410. https://doi.org/10.1287/opre.2.4.393.

[41] M. Held, R.M. Karp, The traveling-salesman problem and minimum spanning trees, Oper. Res. 18 (6) (1970) 1138–1162. https://doi.org/10.1287/opre.18.6.1138.

[42] M. Held, R.M. Karp, The traveling-salesman problem and minimum spanning trees: part II, Math. Program. 1 (1) (1971) 6–25. https://doi.org/10.1007/BF01584070.

[43] S. Kirkpatrick, C.D. Gelatt, M.P. Vecchi, Optimization by simulated annealing, Science (New York, NY) 220 (4598) (1983) 671–680. https://doi.org/10.1126/science.220.4598.671.

[44] J.D.C. Little, K.G. Murty, D.W. Sweeney, C. Karel, An algorithm for the traveling salesman problem, Oper. Res. 11 (6) (1963) 972–989. https://doi.org/10.1287/opre.11.6.972.

[45] P. Miliotis, Using cutting planes to solve the symmetric travelling salesman problem, Math. Program. 15 (1) (1978) 177–188. https://doi.org/10.1007/BF01609016.

[46] J.L. Rodgers, W.A. Nicewater, Thirteen ways to look at the correlation coefficient, Am. Stat. 42 (1) (1988) 59–66.

[47] R. Karp, Reducibility among combinatorial problems, in: R.E. Miller, J.W. Thatcher (Eds.), Complexity of Computer Computations, Plenum, New York, NY, 1972, pp. 85–103.

[48] D.L. Applegate, R.E. Bixby, V. Chvatal, W.J. Cook, The Traveling Salesman Problem: A Computational Study (Princeton Series in Applied Mathematics), Princeton University Press, 2007. Retrieved from http://press.princeton.edu/titles/8451.html.

[49] M. Padberg, G. Rinaldi, Optimization of a 532-city symmetric traveling salesman problem by branch and cut, Oper. Res. Lett. 6 (1) (1987) 1–7. https://doi.org/10.1016/0167-6377(87)90002-2.

[50] A. Boukedjar, M.E. Lalami, D. El-Baz, Parallel branch and bound on a CPU-GPU system, in: 2012 20th Euromicro International Conference on Parallel, Distributed and Network-Based Processing, IEEE, 2012, pp. 392–398. https://doi.org/10.1109/PDP.2012.23.

[51] J. Eckstein, W.E. Hart, C.A. Phillips, PEBBL: an object-oriented framework for scalable parallel branch and bound, Math. Program. Comput. 7 (2015) 429–469. https://doi.org/10.1007/mpc.v0i0.172.

[52] W.J. Cook, In Pursuit of the Traveling Salesman: Mathematics at the Limits of Computation (First), Princeton University Press, 2011. Retrieved from http://press.princeton.edu/titles/9531.html.

[53] S. Climer, W. Zhang, Take a walk and cluster genes: a TSP-based approach to optimal rearrangement clustering, in: Proceedings, Twenty-First International Conference on Machine Learning, ICML 2004, 2004.

[54] M.B. Eisen, P.T. Spellman, P.O. Brown, D. Botstein, Cluster analysis and display of genome-wide expression patterns, Proc. Natl. Acad. Sci. U. S. A. 95 (25) (1998) 14863–14868. Retrieved from http://www.pubmedcentral.nih.gov/articlerender.fcgi?artid=24541&tool=pmcentrez&rendertype=abstract.

[55] M. Newman, M. Girvan, Finding and evaluating community structure in networks, Phy. Rev. E 69 (2) (2004) 026113. https://doi.org/10.1103/PhysRevE.69.026113.

[56] A. Clauset, M. Newman, C. Moore, Finding community structure in very large net-works, Phys. Rev. E 70 (6) (2004) 066111. https://doi.org/10.1103/PhysRevE.70.066111.

[57] J. Duch, A. Arenas, Community detection in complex networks using extremal opti-mization, Phys. Rev. E 72 (2) (2005) 027104. https://doi.org/10.1103/PhysRevE.72.027104.

[58] NEI, Facts About Color Blindness, National Eye Institute, 2019. Retrieved July 11, 2019, from https://nei.nih.gov/health/color_blindness/facts_about.

About the authors

Matthew Lane works as a Software Developer for Bayer Crop Science—St. Louis, MO as well as an Adjunct Instructor at the University of Missouri—St. Louis. His current work at Bayer follows the development of mobile applications for the collection of in-field plant data in remote areas with no existing network connection. His previous works include the building of financial tracking and modeling systems as well as the automation of collection and dis-play of international laws concerning the transport and planting of genetically modi-fied crops and pesticides. Matthew started adjunct teaching Computer Science courses at the University of Missouri—St. Louis in January 2019, and has been volunteer teaching local area high school students basic web development since June 2018.

Alberto Maiocco has worked as an under-graduate research assistant for Dr. Climer at the University of Missouri—St. Louis. His focus is Computer Science and Software Engineering. His interests include machine learning and data mining applications. He currently works for Emerson Electric Co. as an Agile Software Engineer.

Sharlee Climer is an Assistant Professor of Computer Science at the University of Missouri—St. Louis. She received her Ph. D. in Computer Science from Washington University in St. Louis, MO, USA. Her research interests lie at the intersection of Artificial Intelligence and Operations Research and include algorithm development for the extraction of patterns in genetic data and identify associations with traits of interest. Dr. Climer is a member of the team that was awarded ACM's 2018 Gordon Bell Prize, and is a recipient of the Olin and NDSEG Fellowships. She is a faculty member of the Center for Neurodynamics and the Hope Center for Neurological Disorders, and she is the founding director of UMSL's Women CAN organization.

Sanjiv Bhatia works as Professor and Graduate Director (Computer Science) in the University of Missouri—St. Louis. His primary areas of research include Image Databases, Digital Image Processing, and Computer Vision. He has published a number of articles in those areas. He has also consulted extensively with industry for commercial and military applications of computer vision. He is also an expert in system programming and has worked on a few real-time and embedded applications. He serves on the organizing committee of a number of conferences and on the editorial board of a few journals. He has taught a broad range of courses in computer science. He is a senior member of ACM.

CHAPTER FIVE

Eigenvideo for video indexing

Nora Alosily, Sanjiv K. Bhatia

Department of Math & Computer Science, University of Missouri-St. Louis, St. Louis, MO, United States

Contents

Abstract

At present times, the videos are indexed using textual tokens to describe their contents. This works up to an extent but is not very efficient if a user wants to locate a complete video based on a snippet. Further, the user may not describe the video using the exact tokens that have been used for indexing. This leads to our research in indexing based on the contents of a video. Such indexing can also be used to remove redundancy from large libraries of videos. We propose the use of principal component analysis to reduce the dimensionality in large scale video data. This will be further used to determine the best eigenvectors of the dataset that have the smallest eigenvalues. The resultant eigenvectors that we call *eigenvideos*, encapsulate the true dimensionality of video datasets. We present the algorithm that starts by building a representative feature vector for each scene in the dataset. The representative feature vectors are used to determine the eigenvectors that represent the whole dataset. We will present the complete algorithm and show the results using video datasets.

Advances in Computers, Volume 119
ISSN 0065-2458
https://doi.org/10.1016/bs.adcom.2020.03.004

157

1. Introduction

Video libraries and services, such as YouTube, contain a large amount of data and continue to receive even more video data on a daily basis. These libraries are viewed by millions of users globally. Many of those videos may share some scenes, such as trendy clips, recent matches, and interesting news. Such scenes may be of interest to a user who may describe the scene in his own words. Those words form a query to help retrieve the videos that may be of interest to the user. However, such retrieval is feasible and meaningful only if the video has been annotated with the correct text. In addition, the global distribution of users requires that they all use the same language to describe the annotations which may pose a huge problem on its own. Such problems may be overcome if we can provide an example of a clip or scene, and recover a complete video containing that scene. Such a video-based retrieval process overcomes poor annotations and crosses language barriers. Another use of such a system is to identify a complete video that contains a specified scene or clip.

At the most basic level, a video is a sequence of still images, with each image known as a *frame*. Logically speaking, a video is composed of shots and scenes. A *shot* is a sequence of frames that describes a continuous action in time. A shot is delimited by two frames: a starting frame and a terminating frame. A *scene* is a set of semantically correlated shots [1]. The shots and scenes form the *tokens* that can be used to automatically create an index to describe the contents of a video; a technique that is known as *video indexing*.

Video indexing involves solving a multitude of problems, such as indexing the contents of the frames in video, correlating the frames over time, and indexing the action described by those frames over time. The problem is characterized by extremely large data size and information loss. A single video may have multiple shots glued together in several ways; each shot consists of many frames that are basically images of thousands of pixels, with a typical video recorded at 30 frames per second (FPS). Unlike text indexing, video indexing tends to deal with information loss in different levels and measures. A video may lose some shots or some frames in a shot; the whole video may be resized; and pixel values may be shifted while applying filters. However, the human eye can easily identify different videos with such loss as containing identical material, and sometime, may not even recognize the change. Thus, indexing videos automatically may lead to loss of information contained therein.

A video is characterized by the information contained in its frames (spatial component) as well as an action given by a shot over multiple frames (temporal component). A user may want to frame a query on either of those components. For example, the user may want to retrieve a video containing a ship, or he may want to retrieve the video containing a man kicking a ball. While a video by itself contains a very large amount of raw spatial data due to the images constituting the frames, the addition of a semantic or temporal component adds yet another dimension of complexity in the representation of videos, further increasing the storage requirements.

The large amount of data may be rendered manageable by some techniques such as principal component analysis (PCA). PCA is a statistical technique used for dimensionality reduction and extraction of relevant data. The approach aims to capture the dimensions that have the greatest variance [2]. The procedure is lossy but works especially well with high dimensional and correlated data. These characteristics are well suited to video indexing. We propose a way to apply PCA to index video shots, in a fashion similar to its usage in the well-known Eigenfaces algorithm [3–5], as a system of multiple models represented by their average shot. We call our structures as *eigenvideo*.

In this chapter, we present the use of PCA to index videos. We will present the algorithms for indexing and retrieval of video data, and show the effectiveness of the technique experimentally. In the next section, we will present some of the related work. This is followed by a brief presentation of the data set used in our experiments. In Section 4, we will present the use of PCA to index videos and use the index to perform retrieval. In Section 5, we will demonstrate the use of our techniques using standard publicly available data sets, followed by conclusion.

2. Background

In this section, we will look at different approaches used to extract signatures and perform indexing. We will discuss different ways to analyze videos including shot analysis, keyframes, generation of signatures for videos, and indexing of the generated signatures.

Some of the early work on video indexing was reported by Smoliar and Zhang [6]. They identified the problem solution into three subproblems as parsing, indexing, and retrieval. The step of parsing, or splitting a video into clips, is important because it specifies the granularity of the index. If each chunk is a single frame, signatures will be extracted from each frame, increasing the computational and spatial complexity. Subsequently, the final retrieval would be based on the signatures from all frames. If the chunk is a whole video, each video will be represented with a single signature, and retrieval will be done using the rough global signature, reducing the computations substantially [7]. A way in between is to split the video into shots and extract signatures from those shots. The final retrieval will be based on comparing signatures of the shots. Once parsing is done, signatures are then extracted, possibly organized into some structure for efficient access. Finally, when receiving a query video, the index is extracted in similar fashion, and the built index is searched, and best matches are retrieved.

More recently, Liu et al. [7] have suggested four levels at which signatures are extracted: frame-local signatures, frame-global signatures, global signatures, and spatiotemporal signatures. Local signatures are ones that are extracted from each frame in the video, increasing the computational and spatial complexity. Frame-global signatures represent the whole frame with a single signature, which enhances both the aforementioned complexities. In both cases, videos are compared based on signatures from each frame. Global signatures represent the whole video with a single signature, so videos are compared globally. Finally, spatiotemporal signatures include spatial and temporal information. Shot analysis can fit in both the frame level and spatiotemporal signatures based on the choice of shots.

In the remainder of this section, we will look at different ways of analyzing videos to determine their signature.

2.1 Shot analysis and keyframes

In this section, we will look at the methods to analyze and extract signatures from shots. At the shot level, there are three types of analysis: keyframe analysis, object analysis, and motion analysis. Keyframes work as representatives of the entire shot and may be identified as a single frame or set of multiple frames [8]. The keyframe selection can be done with various methods, such as comparison between frames [9], comparison to a reference frame, and frame clustering [10]. Once keyframes are obtained, either low-level features or high-level features are extracted as the shot signatures. The low-level features include color, texture, and shape features. Here, a whole shot may be represented with texture descriptors extracted from its keyframes. On the other hand, high-level features include objects and concepts [11, 12]. That is, a shot can be represented with certain objects extracted from its keyframes, such as faces or cars. Object analysis is very similar to the latter, but each frame in the shot is processed to extract target objects as signatures. Finally, for shot motion analysis, camera-based and object-based motions are used to extract different types of motion-based signatures [13].

Object analysis lacks the spatial information used to model the arrangement of objects in space, as well as the temporal information that arranges them in time. In contrast, motion analysis usually focuses on the trajectory regardless of the identity of the moving object and other features of the object. Thus, keyframe analysis is interesting in that, except for motion, it can take into account the background and foreground objects, including low-level features such as edges, color, and texture. However, there are two problems that arise when selecting a subset of frames: the difficulty to reproduce the exact scene boundaries, and the absence (or addition) of features in the objects in frames with respect to the reference frame. The first problem is encountered in case of shot-to-shot retrieval. Due to possible loss in visual content, there must be no assumption of perfect shot boundary matching between those in a query clip and its matching indexed video. Thus, omitting or adding some frames should not affect the result. Missing a frame may turn problematic if that frame happens to be a keyframe in the indexed shot. Additionally, many methods allow variable number of selected keyframes which is hard to reproduce with a given query. The second problem turns up when some features like motion are omitted. Motion is an essential feature to video and makes it unique compared to an arbitrary image collection. The lack of motion makes video indexing closer to image indexing problem. Thus, we need a feature vector that encompasses the whole shot in a meaningful way. Those vectors will be used as the seeds to build the whole indexing system. In Section 4.1, we propose an efficient method to extract such feature vector F that is linear to the number of frames in a shot.

2.2 Signatures

Once features like keyframes are extracted, we need a way to index and access the features contained therein in an efficient manner. This leads to the problems of content-based video retrieval (CBVR) and near duplicate video retrieval (NDVR) which are very close problems. They both index videos from signatures extracted using visual content rather than other types of data, like subtitles. While NDVR allows a broader definition of what makes videos duplicates, CBVR applies a more restricted definition of similarity. Moreover, a query in NDVR is some type of video, while it can be of any form in case of CBVR. However, retrieval is based on the visual content in both cases. Both have similar framework for indexing and retrieval as shown in Fig. 1B. For elaborated reviews on CBVR and NDVR, refer to [7, 14–16].

Semantic concepts, such as walking, crowd, and overlaid text that are inferred from video content, can be used as video signatures. Liu et al. [17] have applied association rule mining to improve the semantic concept detection. Naphade and Hang [18] exploit Bayesian networks to infer semantic concepts from the visual content. Weng and Chuang [19] propose a method to infer contextual and temporal relationships constructed for each concept in order to predict concepts from unseen shots.

Signatures can also be inferred from objects contained within the video frames. The index can be built by first detecting the objects, and then using those to build signatures. During retrieval, search can be performed using signatures from objects detected in a query shot. Gao and Yang [20] index videos based on features from their salient objects. Each object is represented with its low-level features, spatial features, and its temporal trajectory, and the extracted features are used to perform comparisons. Faces as objects are also used to index shots [21]. Once detected and tracked, descriptors from faces are used to index shots.

Another technique to generate signatures is based on hashing. Given a video, the hashing algorithm produces a signature vector (hash), that is replicable when retrieving an identical query. De Roover et al. [22] have investigated the use of discrete cosine transform (DCT) coefficients to represent the hash signatures. Those coefficients are extracted from radial projections of keyframes, and retrieval is done based on the similarity of DCT coefficients. Hashes may also be extracted from multiple features, like local and spatiotemporal features in keyframes [23]. Song et al. [24] construct an autoencoder out of several LSTMs (long short-term memory) networks. In the architecture, several LSTMs, a type of neural network that is able to learn sequential patterns, are used in each part of the autoencoder structure to generate the hashes.

Another concept utilized by some authors to generate signatures is based on dimensionality reduction techniques. Dimensionality reduction methods such as PCA have been used to generate signatures by projecting data into a smaller space. A famous application of PCA is the Eigenfaces algorithm for face recognition [3]. The algorithm extracts the desired number of principal components, the Eigenfaces, and projects the dataset onto the new space.

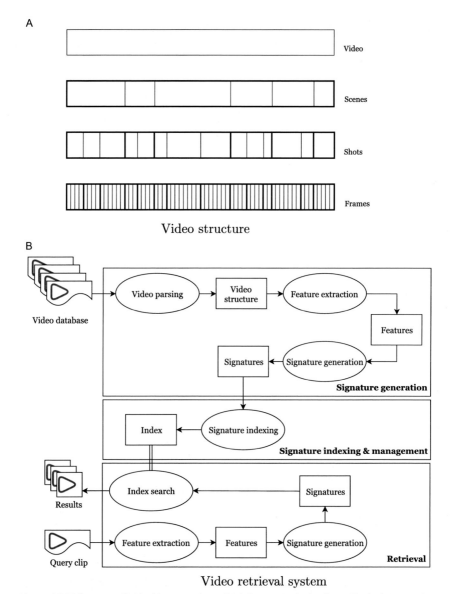

Video structure

Video retrieval system

Fig. 1 (A) Videos are divided into scenes, which in turn include shots. Each shot consists of several frames. (B) Once the structure has been decided in the parsing phase, features are extracted from that structure. Signatures from those features are generated. Signatures are then indexed and managed for more efficient access. Retrieval is shown for shot queries as an example.

For prediction, the algorithm predicts the class of a new observation based on the distance between its projection and the rest of the dataset projections. Related algorithms that followed include Fisherfaces [25] that utilizes linear discriminant analysis (LDA), in which the dimensionality reduction is class-specific. In this chapter, we use the standard PCA and Eigenfaces algorithm to extract small signatures from larger shots features.

Next, we will look at building an index using the signature developed for the video.

2.3 Signature indexing

We build an index to perform search in a more efficient and effective manner. The use of an index helps in reducing the time to look for an object of interest. We are familiar with text index, for example in a book, that helps us get to a chapter or section much faster than searching for it brute force. Text indexing is a well-studied problem, with a plethora of methods addressing it. With the proliferation of online texts, there have been new techniques to build indices such as Bag-of-Words (BoW) [26]. BoW is a way to represent a text document by the occurrence of words in it rather than their order or semantics. In text indexing, the model can be used as an early step toward more structured and efficient indexing schemes. The BoW concept has been applied in computer vision context successfully with various interpretations of visual words, where visual words are used as a reference to low-level features in an image or a frame.

Since BoW does not deal with the order of words, it is replaced or accompanied by methods that carry out the spatial flow component for videos. Similar concept can be applied in video indexing to deal with the temporal ordering of shots. Visual *sentence representation* introduced in [26] addresses the spatial flow problem in images with great results. Having shots reduced to small signatures as described above, applying visual sentence representation is feasible at this point.

With the above background in mind, we are ready to describe our algorithm for video indexing. In the next section, we will describe the dataset used to test our algorithm followed by the algorithm itself.

3. Dataset

We use KTH action dataset [27], a widely used dataset for human action recognition to test our indexing and retrieval algorithm. The dataset is made up of a set of six actions (boxing, clapping, jogging, running, waving, and walking). Each action is performed by 25 persons in four different scenarios (outdoors, outdoors with scale variation, outdoors with different clothes, and indoors). The resolution of video sequences is 160×120 pixels, with pre-specified shot boundaries resulting in 2391 shots.

We selected this dataset for two reasons. First, it contains the variability needed to evaluate a video-based indexing method. Certain actions are very similar, like jogging and running, while others are very disparate like clapping and walking. Certain shots share the same background but differ in action and person performing, while others share the action and differ in the scenario. Second, although it is rich in content, the dataset is simple in terms of size and dimensions compared to most of the popular datasets currently available. That makes it work as a toy example for the indexing method discussed here.

4. Algorithms

In this section, we describe our proposed algorithms for shot-based video indexing, from shot feature extraction through eigenvideo extraction to final shot retrieval. The overall process is presented in Fig. 2. We first describe shot feature extraction in Section 4.1. Then, in Section 4.2, we explain in details the use of PCA in a manner similar to Eigenfaces algorithm [3] but for video shots. Finally, we show the process to retrieve indexed shots in Section 4.3.

4.1 Extraction of shot features

Given a sequence of frames in a shot where boundaries begin from *start* and terminate at *end*, we need to extract a feature vector that represents the

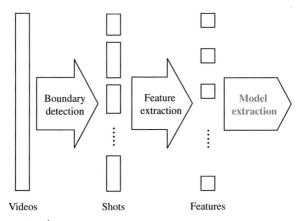

Fig. 2 Feature extraction.

whole shot. The frames in a shot are of size $d = R \times C$ pixels where R and C denote the number of rows and columns, respectively, in each frame of the shot. d also denotes the length of the feature vector F. Initially, we compute F_B to represent the *background* of a shot (the static component) as the summation of all corresponding pixels in each frame f_k in the shot:

$$F_B = \sum_{k=\text{start}}^{\text{end}} f_k \tag{1}$$

After computing F_B, we apply feature scaling, namely min–max normalization, with values ranging from α to β, where α and β describe the range of pixel intensities in the frames or video. That is, each value v_i in F is computed using Eq. (2). Normalization is important because of the need to avoid bias toward long video shots. Also, the normalization makes it appropriate to use covariance rather than the correlation for better complexity.

$$v_i = \alpha \frac{(v_i - F_{\min}) \times (\beta - \alpha)}{F_{\max} - F_{\min}} \tag{2}$$

After computing the background, we need another vector F_M that represents the motion in the shot. To extract motion, we apply an approach similar to Eq. (1), but with the difference vector:

$$F_M = \sum_{k=\text{start} + 1}^{\text{end}} f_k - f_{k-1} \tag{3}$$

Finally, we combine both vectors to get the feature vector that describes both motion and its background in a given shot.

$$F = F_B + F_M \tag{4}$$

We provide examples of some frames in a given shot, and their corresponding feature vector in Fig. 3, where we chose $\alpha = 0$ and $\beta = 255$ (for 8-bit grayscale frames). The overall process to compute feature vectors is described in Algorithm 1.

In the next section, we provide details of principal component analysis to extract eigenvideo.

4.2 Principal component analysis (PCA)

Let S be the set of n feature vectors in the dataset, such that $S = [F_1, F_2, ..., F_n]$, and each feature vector F is of d dimensions. We need to find the best k eigenvectors with the largest variance that can express the whole dataset, such that $k \ll d$. Thus, each feature vector can be represented

Fig. 3 Samples of frames in certain shots, along with their corresponding background vector (F_B), motion vector (F_M), and feature vector (F).

using only k values instead of d values. All steps are shown in Algorithm 2 and in Fig. 4. Some of the important points are explained in Sections 4.2.1–4.2.5.

4.2.1 Mean μ and centered samples A

While PCA can be computed without centering data around the mean [2], mean centering is needed for covariance matrix calculation and better extraction of the eigenvectors. Additionally, we will use the average vector μ for retrieval and as a representative of each model. We first calculate the mean μ of the set $S = [F_1, F_2, ..., F_n]$ as follows:

$$\mu = \frac{1}{n} \sum_{i=1}^{n} F_i \tag{5}$$

ALGORITHM 1 Shot-feature-extraction.

    input : A single shot S where each frame has d dimensions
           Shot's boundaries given by *start* and *end*
    output: Feature vector $F = \{f_1, ..., f_d\}$

    F_B is the background vector and F_M is the motion vector
1   $F_B \leftarrow 0$
2   $F_M \leftarrow 0$

    Calculate the summation of d dimension in each frame f_i
3 **for** $i \leftarrow$ *start* **to** *end* $- 1$ **do**
4   │ $F_B \leftarrow F_B + f_i$
5   │ $F_M \leftarrow F_M + (f_{i+1} - f_i)$
6 **end**

7 $F \leftarrow$ Normalize(F_B, α, β) + Normalize(F_M, α, β)

Then, we center the set S around the mean. The centered S, is now $A = [a_1, a_2, ..., a_n]$, such that $a_i = F_i - \mu$. The size of matrix A is $d \times n$.

4.2.2 Covariance matrix C

To find the correlation between the different dimensions d in the n samples, we first calculate the covariance or correlation matrix. The correlation matrix is needed when dimensions are not of unit length [2]. Since the feature vectors are normalized, the covariance matrix C is used and calculated as follows:

$$C = \frac{1}{n} \sum_{i=1}^{n} A \, A^T \tag{6}$$

4.2.3 Eigenvectors

For a matrix of size $d \times n$, there are only $\min\{d, n\}$ unique eigenvectors.

$$Cu_i = \lambda_i u_i, \quad i = 1, 2, ..., \min\{d, n\} \tag{7}$$

4.2.4 Principal components (Eigenvideo)

The principal components are a subset of eigenvectors with the largest eigenvalues. By ordering the eigenvalues in a nonincreasing order $\lambda_1 \geq \lambda_2 \geq \cdots \geq \lambda_{\min\{d,n\}}$, we can pick the first corresponding k eigenvectors

ALGORITHM 2 Create-single-model.

input : Set of n shots features $S = \{F_1, ..., F_n\}$, each $F_i = [v_1, ..., v_d]^T$
Number of desired principal components k

output: Mean of the model μ
Set of k eigenvectors $U_k = \{u_1, ..., u_k\}$
Set of n projections for each feature into the new space k,
$\Omega = \{\Omega_1, ..., \Omega_n\}$, each $\Omega_i = [\omega_1, ..., \omega_k]^T$

Calculate the average feature vector μ
1 **for** $i \leftarrow 1$ **to** n **do**
2 $\quad | \quad \mu \leftarrow F_i + \mu$
3 **end**
4 $\mu \leftarrow \frac{\mu}{m}$

Calculate the mean-centered vector a_i for each feature vector F_i
5 **for** $i \leftarrow 1$ **to** n **do**
6 $\quad | \quad a_i \leftarrow F_i - \mu$
7 **end**
8 Combine $[a_1, ..., a_n]$ into matrix $A_{d \times n}$

*Find the covariance matrix $C = AA^T$, eigenvectors $U = [u_1, ..., u_n]$
and their eigenvalues $\Lambda = [\lambda_1, ..., \lambda_n]$*
9 **if** $d \leq n$ **then**
10 $\quad |$ Calculate using PCA, such that $AA^T U = \Lambda U$
11 **else**
12 $\quad |$ Calculate as in Eigenfaces, such that $AA^T AU' = \Lambda AU'$
13 $\quad | \quad U = AU'$
14 **end**

Extract k principal components
15 $U_k \leftarrow$ Pick k eigenvectors with the highest eigenvalues, such that $k << d$

Project each feature vector F_i into the new space k
16 **for** $i \leftarrow 1$ **to** n **do**
17 $\quad | \quad t_i \leftarrow$ Project-Into-K-Space(a_i, U_k)
18 **end**

(eigenvideos) as the principal components $U_k = [u_1, u_2, ..., u_k]$. Matrix U_k combines all the eigenvideos and is of size $d \times k$.

4.2.5 Projection

Once the k eigenvideos of the model are computed, we project each vector a_i into k-space using the following equation:

$$T = A^T U_k \qquad (8)$$

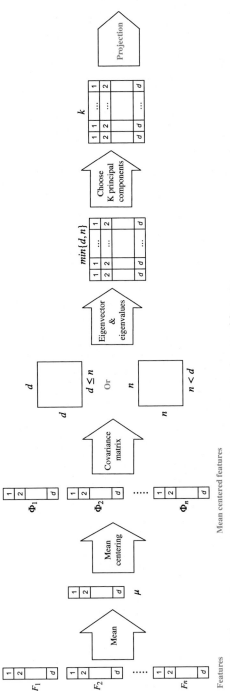

Fig. 4 Model extraction: extracting *k* principal components out of a set of features.

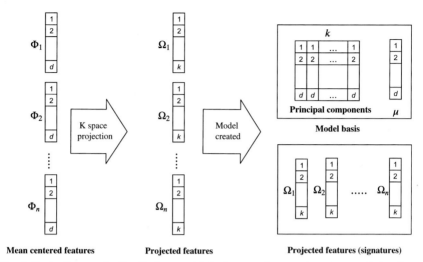

Fig. 5 Projection: projecting features set into the new k-space to extract only k values for each feature instead of d values.

The resulting matrix $T = [t_1, t_2, \ldots, t_n]^T$ is of size $n \times k$. Each vector t_i is the projection of its corresponding a_i in the new space k instead of d. We show the projection and model extraction in Fig. 5.

4.3 Retrieval

To retrieve a shot, we simply extract its feature vector and project the mean-centered vector into the space k, using the method described above. Finally, we find its nearest match by computing the distance between the projected query and the projected data.

5. Experiments

We have tested the eigenvideo algorithm in two ways, once using a single model and then, using two models for indexing and retrieval. The two models example demonstrates building a simple system where the leaf nodes are the projected shots, and the nonleaf nodes are the models that contain the mean vectors for mapping and the eigenvectors for projecting. We believe that building such a system of many models is a good start for video to video retrieval. All queries are shots and so is the indexed content.

5.1 Single model

In this section, we provide numerical results illustrating the performance of PCA models having different number of principal components. Initially, we perform the experiments on the whole dataset without any clustering. The approach does not fully utilize the capability of regular PCA to capture principal components in similar samples. Hence, the principal components in this case capture the differences between classes rather than the differences within the same class. We evaluate the results using the standard measures of precision and recall. Precision is the ratio of the number of true positives in the retrieved set to the number of instances in the retrieved set. Recall is the ratio of true positives in the retrieved set to the total number of true positives in the collection.

In Tables 1–3, we show results of models with variable number of components. The results can be visualized in Fig. 6. Each run is performed using 2391 queries, which are the total number of indexed shots. The number of relevant shots is always four, considering there is only four shots that have the same person, action and setting. Also, we perform the queries using both L1 and L2 as similarity measures. The compression of a single model reduces the size of the dataset by around 99%[a] which can be more apparent when dealing with much larger datasets.

5.2 Multiple models

The eigenvideo algorithm proposed in this chapter is intended to work for systems with multiple models, each having similar shots. Hence, we performed an experiment with a simple system of two models. We used the k-means algorithm to cluster the feature vectors F into two groups of shots. Then, we built a model for each cluster. Finally, when performing retrieval,

Table 1 Precision and recall of top k queries in a model with 10 principal components.

	Recall		Precision	
K	L1	L2	L1	L2
1	0.25	0.25	1	1
2	0.41698	0.409557	0.833961	0.819113
4	0.655583	0.622857	0.655583	0.622857
6	0.727834	0.692493	0.485222	0.461662
8	0.775931	0.736198	0.387965	0.368099
10	0.808762	0.764743	0.323505	0.305897

---

[a] Data size reduction from 1162 MB to about 6.5 MB.

Table 2 Precision and recall of top k queries in a model with 15 principal components.

	Recall		Precision	
K	L1	L2	L1	L2
1	0.25	0.25	1	1
2	0.431305	0.422836	0.86261	0.845671
4	0.700439	0.661857	0.700439	0.661857
6	0.776872	0.731702	0.517914	0.487801
8	0.821936	0.778754	0.410968	0.389377
10	0.85184	0.802907	0.340736	0.321163

Table 3 Precision and recall of top k queries in a model with 20 principal components.

	Recall		Precision	
K	L1	L2	L1	L2
1	0.25	0.25	1	1
2	0.440715	0.428586	0.88143	0.857173
4	0.722396	0.681409	0.722396	0.681409
6	0.796738	0.752719	0.531159	0.501812
8	0.842744	0.795169	0.421372	0.397585
10	0.871602	0.822355	0.348641	0.328942

we first selected the cluster that is most similar to our query based on the centroid of that cluster. The rest of the retrieval process is identical to the single model algorithm presented in the last subsection.

The results are presented in Tables 4-6 and visualized in Fig. 7.

While some methods referenced in the background require a prior knowledge to identify and build objects or concepts, this method is simple and reduces the size without any prior knowledge. Moreover, all other methods try to analyze the data at video level and let the signatures be the last step, we can do opposite. That is, after building the models, we can analyze at model level. Hence multiple indexing schemes can be built out of this simple method at the same time.

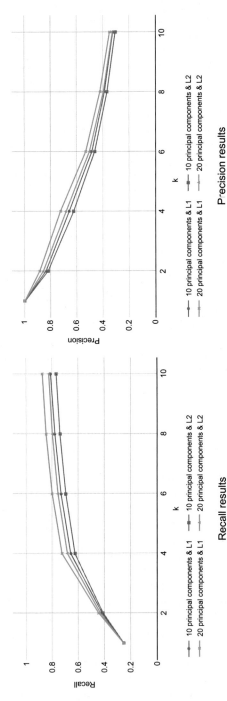

Fig. 6 Recall and precision with various number of principal components of a system with a single model.

Table 4 Precision and recall of top k queries in a model with 10 principal components from two models.

	Recall		Precision	
K	$L1$	$L2$	$L1$	$L2$
1	0.25	0.25	1	1
2	0.413216	0.406211	0.826432	0.812422
4	0.638018	0.614283	0.638018	0.614283
6	0.707445	0.678586	0.47163	0.452391
8	0.750418	0.719155	0.375209	0.359578
10	0.774676	0.74519	0.30987	0.298076

Table 5 Precision and recall of top k queries in a model with 15 principal components from two models.

	Recall		Precision	
K	$L1$	$L2$	$L1$	$L2$
1	0.25	0.25	1	1
2	0.428586	0.419594	0.857173	0.839189
4	0.679946	0.647846	0.679946	0.647846
6	0.750209	0.714032	0.500139	0.476021
8	0.793078	0.753973	0.396539	0.376987
10	0.815454	0.777394	0.326182	0.310958

Table 6 Precision and recall of top k queries in a model with 20 principal components from two models.

	Recall		Precision	
K	$L1$	$L2$	$L1$	$L2$
1	0.25	0.25	1	1
2	0.434442	0.425136	0.868883	0.850272
4	0.696884	0.664994	0.696884	0.664994
6	0.765161	0.729506	0.510107	0.486338
8	0.805102	0.76903	0.402551	0.384515
10	0.828837	0.79496	0.331535	0.317984

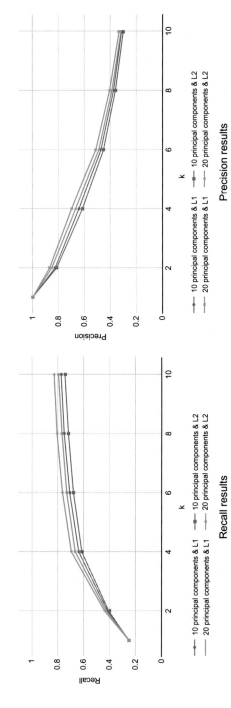

Fig. 7 Recall and precision with various number of principal components of system with two models.

6. Conclusion and future work

In this chapter, we have presented the building blocks for a video indexing system. We have presented a way to index the videos using PCA for video retrieval, annotation, and classification. We have performed experiments using a single model and a two models architecture for indexing and retrieval.

Along the way, we have faced a few issues. We have been hampered a bit by the lack of a very large dataset to test our system. It is expected that with the growth of data in the system, the models will be more expressive, and the compression of the data structures will be more pronounced. We are also investigating a better way to cluster shots in order to build the models.

At present time, we have been using comparison of the projection of the query to all other projections in the nearest model. We are investigating the use of some hashing techniques or some more efficient trees for faster retrieval. We find the hashing model interesting since the number of dimensions k in the new spaces is limited, as in 10–20 principal components.

In future, we plan to improve extracting the feature vector F by replacing F_M with motion history image [28]. Queries will also be whole videos or at least several consecutive shots, and we expect higher accuracy without introducing much complexity using sentence representation. We also intend to test the method with larger datasets. Furthermore, we plan to look into K-D trees for a faster and partial retrieval.

References

[1] T. Lin, H.-J. Zhang, Automatic video scene extraction by shot grouping, in: ICPR 2000: Proceedings of the 15th International Conference on Pattern Recognition, vol. 42000, pp. 39–42.

[2] I.T. Jolliffe, J. Cadima, Principal component analysis: a review and recent developments, Philos. Trans. R. Soc. A Math. Phys. Eng. Sci. 374 (2065) (2016) 20150202.

[3] M. Turk, A. Pentland, Eigenfaces for recognition, J. Cogn. Neurosci. 3 (1) (1991) 71–86.

[4] M.A. Turk, A.P. Pentland, Face recognition using eigenfaces, in: CVPR '91: Proceedings of the IEEE Computer Society Conference on Computer Vision and Pattern Recognition, June, HI, Maui, 1991, pp. 586–591.

[5] A. Pentland, T. Starner, N. Etcoff, A. Masoiu, O. Oloyide, M. Turk, Experiments with Eigenfaces, in: Looking at People Workshop, IJCAI '93, August, 1993, Chamberry, France.

[6] S.W. Smoliar, H. Zhang, Content-based video indexing and retrieval, IEEE Multimedia 1 (2) (1994) 62–72.

[7] J. Liu, Z. Huang, H. Cai, H.T. Shen, C.W. Ngo, W. Wang, Near-duplicate video retrieval: current research and future trends, ACM Comput. Surv. (CSUR) 45 (4) (2013) 44.

[8] Y. Rui, T.S. Huang, S. Mehrotra, Exploring video structure beyond the shots, in: Proceedings of the IEEE International Conference on Multimedia Computing and Systems (Cat. No. 98TB100241), IEEE, 1998, pp. 237–240.

[9] X.-D. Zhang, T.-Y. Liu, K.-T. Lo, J. Feng, Dynamic selection and effective compression of key frames for video abstraction, Pattern Recogn. Lett. 24 (9–10) (2003) 1523–1532.

[10] A. Girgensohn, J. Boreczky, Time-constrained keyframe selection technique, Multimed. Tools Appl. 11 (3) (2000) 347–358.

[11] A. Hampapur, R. Jain, T.E. Weymouth, Feature based digital video indexing, in: Working Conference on Visual Database Systems, Springer, 1995, pp. 115–141.

[12] Y. Liu, D. Zhang, G. Lu, W.-Y. Ma, A survey of content-based image retrieval with high-level semantics, Pattern Recogn. 40 (1) (2007) 262–282.

[13] Y.-F. Ma, H.-J. Zhang, Motion texture: a new motion based video representation, in: Object Recognition Supported by User Interaction for Service Robots, vol. 2, IEEE, 2002, pp. 548–551.

[14] W. Hu, N. Xie, L. Li, X. Zeng, S. Maybank, A survey on visual content-based video indexing and retrieval, IEEE Trans. Syst. Man Cybern. C: Appl. Rev. 41 (6) (2011) 797–819.

[15] F. Idris, S. Panchanathan, Review of image and video indexing techniques, J. Vis. Commun. Image Represent. 8 (2) (1997) 146–166.

[16] V. Vijayakumar, R. Nedunchezhian, A study on video data mining, Int. J. Multimed. Inf. Retr. 1 (3) (2012) 153–172.

[17] K.-H. Liu, M.-F. Weng, C.-Y. Tseng, Y.-Y. Chuang, M.-S. Chen, Association and temporal rule mining for post-filtering of semantic concept detection in video, IEEE Trans. Multimed. 10 (2) (2008) 240–251.

[18] M.R. Naphade, T.S. Huang, Semantic video indexing using a probabilistic framework, in: Proceedings 15th International Conference on Pattern Recognition. ICPR-2000, vol. 3 IEEE, 2000, pp. 79–84.

[19] M.-F. Weng, Y.-Y. Chuang, Multi-cue fusion for semantic video indexing, in: Proceedings of the 16th ACM International Conference on Multimedia, ACM, 2008, pp. 71–80.

[20] H.-P. Gao, Z.-Q. Yang, Content based video retrieval using spatiotemporal salient objects, in: 2010 International Symposium on Intelligence Information Processing and Trusted Computing, IEEE, 2010, pp. 689–692.

[21] J. Sivic, M. Everingham, A. Zisserman, Person spotting: video shot retrieval for face sets, in: International Conference on Image and Video Retrieval, Springer, 2005, pp. 226–236.

[22] C. De Roover, C. De Vleeschouwer, F. Lefebvre, B. Macq, Robust video hashing based on radial projections of key frames, IEEE Trans. Signal Process 53 (10) (2005) 4020–4037.

[23] J. Song, Y. Yang, Z. Huang, H.T. Shen, R. Hong, Multiple feature hashing for real-time large scale near-duplicate video retrieval, in: Proceedings of the 19th ACM International Conference on Multimedia, ACM, 2011, pp. 423–432.

[24] J. Song, H. Zhang, X. Li, L. Gao, M. Wang, R. Hong, Self-supervised video hashing with hierarchical binary auto-encoder, IEEE Trans. Image Process. 27 (7) (2018) 3210–3221.

[25] P.N. Belhumeur, J.P. Hespanha, D.J. Kriegman, Eigenfaces vs. fisherfaces: recognition using class specific linear projection, IEEE Trans. Pattern Anal. Mach. Intell. 19 (7) (1997) 711–720.

[26] P. Tirilly, V. Claveau, P. Gros, Language modeling for bag-of-visual words image categorization, in: Proceedings of the 2008 International Conference on Content-based Image and Video Retrieval, ACM, 2008, pp. 249–258.

[27] C. Schüldt, I. Laptev, B. Caputo, Recognizing human actions: a local SVM approach, in: ICPR 2004: Proceedings of the 17th International Conference on Pattern Recognition, IEEE, 2004, pp. 32–36.

[28] J.W. Davis, A.F. Bobick, The representation and recognition of action using temporal templates, in: IEEE Conference on Computer Vision and Pattern Recognition, 1997, pp. 928–934.

About the authors

Nora Alosily received BS in Computer Science from Qassim University, Saudi Arabia, where she worked as a teaching assistant. She received MS in Computer Science from University of Missouri—St. Louis in 2017. She is currently a PhD candidate in Mathematical and Computational Sciences at the University of Missouri-St. Louis, with focus on computer vision. Her interests include algorithms and machine learning.

Sanjiv Bhatia works as professor and graduate director (Computer Science) in the University of Missouri—St. Louis. His primary areas of research include Image Databases, Digital Image Processing, and Computer Vision. He has published a number of articles in those areas. He has consulted extensively with industry for commercial and military applications of computer vision. He is also an expert in system programming and has worked on a few real-time and embedded applications. He serves on the organizing committee of a number of conferences and on the editorial board of a few journals. He has taught a broad range of courses in computer science. He is a senior member of ACM.

Printed in the United States
By Bookmasters